Communications
in Computer and Information Science　　2158

AF173230

Rationale

The CCIS series is devoted to the publication of proceedings of computer science conferences. Its aim is to efficiently disseminate original research results in informatics in printed and electronic form. While the focus is on publication of peer-reviewed full papers presenting mature work, inclusion of reviewed short papers reporting on work in progress is welcome, too. Besides globally relevant meetings with internationally representative program committees guaranteeing a strict peer-reviewing and paper selection process, conferences run by societies or of high regional or national relevance are also considered for publication.

Topics

The topical scope of CCIS spans the entire spectrum of informatics ranging from foundational topics in the theory of computing to information and communications science and technology and a broad variety of interdisciplinary application fields.

Information for Volume Editors and Authors

Publication in CCIS is free of charge. No royalties are paid, however, we offer registered conference participants temporary free access to the online version of the conference proceedings on SpringerLink (http://link.springer.com) by means of an http referrer from the conference website and/or a number of complimentary printed copies, as specified in the official acceptance email of the event.

CCIS proceedings can be published in time for distribution at conferences or as postproceedings, and delivered in the form of printed books and/or electronically as USBs and/or e-content licenses for accessing proceedings at SpringerLink. Furthermore, CCIS proceedings are included in the CCIS electronic book series hosted in the SpringerLink digital library at http://link.springer.com/bookseries/7899. Conferences publishing in CCIS are allowed to use Online Conference Service (OCS) for managing the whole proceedings lifecycle (from submission and reviewing to preparing for publication) free of charge.

Publication process

The language of publication is exclusively English. Authors publishing in CCIS have to sign the Springer CCIS copyright transfer form, however, they are free to use their material published in CCIS for substantially changed, more elaborate subsequent publications elsewhere. For the preparation of the camera-ready papers/files, authors have to strictly adhere to the Springer CCIS Authors' Instructions and are strongly encouraged to use the CCIS LaTeX style files or templates.

Abstracting/Indexing

CCIS is abstracted/indexed in DBLP, Google Scholar, EI-Compendex, Mathematical Reviews, SCImago, Scopus. CCIS volumes are also submitted for the inclusion in ISI Proceedings.

How to start

To start the evaluation of your proposal for inclusion in the CCIS series, please send an e-mail to ccis@springer.com.

Henry Han
Editor

Recent Advances in Next-Generation Data Science

Third Southwest Data Science Conference, SDSC 2024
Waco, TX, USA, March 22, 2024
Revised Selected Papers

 Springer

Editor
Henry Han (ID)
Baylor University
Waco, TX, USA

ISSN 1865-0929 ISSN 1865-0937 (electronic)
Communications in Computer and Information Science
ISBN 978-3-031-67870-7 ISBN 978-3-031-67871-4 (eBook)
https://doi.org/10.1007/978-3-031-67871-4

Preface

The rapid evolution of AI and the emergence of cutting-edge computing technologies, such as quantum and neuromorphic computing, have fundamentally transformed the landscape of next-generation data science, enabling unprecedented capabilities and automation. AI technologies, including machine learning (ML) and deep learning, have revolutionized data science by enhancing the ability to process and analyze large datasets with remarkable accuracy and speed. These technologies are employed in diverse applications such as healthcare diagnostics, financial forecasting, autonomous vehicles, and natural language processing (NLP). The integration of AI has not only improved efficiency and decision-making but also opened new avenues for innovation and discovery.

However, as AI becomes increasingly integral to, or even dominant in, critical applications across various data science domains, ensuring its security has emerged as a paramount concern. AI security in next-generation data science involves addressing a range of challenges, from protecting sensitive data to mitigating adversarial threats. To some degree, the security of next-generation data science is synonymous with AI security in data science.

One of the primary security challenges in AI-driven next-generation data science is data privacy. AI systems often require vast amounts of data for training and optimization, much of which can be sensitive or personally identifiable information (PII). Ensuring that this data is protected from unauthorized access and breaches is critical. Techniques such as differential privacy, which adds noise to data to obscure individual information while preserving overall trends, are essential in mitigating privacy risks.

Adversarial attacks pose another significant threat to AI security in data science. These attacks involve manipulating input data to deceive AI models into making incorrect predictions or classifications. For instance, in the fintech industry, AI models are often used to detect fraudulent transactions by analyzing various features of a transaction, such as the amount, location, time, and the history of the user's transactions, to determine the likelihood of fraud. Adversarial examples can be crafted by initiating transactions at times that align with the user's usual activity patterns, thereby evading time-based anomaly detection. Such vulnerabilities can have serious consequences in other data science domains, especially in safety-critical applications like autonomous driving or medical diagnostics.

AI model integrity is another critical aspect of AI security in data science. AI models can be susceptible to attacks that alter their internal parameters or logic, leading to compromised performance. For example, introducing biased or malicious data during the training phase can corrupt the model's learning process, resulting in biased or incorrect outputs and changes to parameter settings. In some cases, AI models can be biased due to artifacts in training and deployment.

Enhancing the robustness of AI models in data science against adversarial attacks requires ongoing research and development. Techniques such as adversarial training, where models are trained on adversarial examples to improve their resilience, and defensive distillation, which reduces the model's sensitivity to small input perturbations, can bolster model security. Additionally, employing AI-driven monitoring systems that continuously analyze model performance and detect anomalies can help identify and mitigate potential threats in real time, especially for high-stakes data science applications.

Moreover, transparency and explainability are crucial components of AI security in data science to enhance AI model integrity. Ensuring that AI models are interpretable and their decision-making processes are transparent can help identify biases, errors, and vulnerabilities. Explainable AI (XAI) techniques enable stakeholders to understand and trust AI systems, promoting accountability and facilitating the detection of malicious activities. Furthermore, investment in data explainability will provide a novel perspective to enhance XAI, as it can be superficial to discuss a model's explainability without considering the explainability of the data.

Emerging technologies like blockchain and quantum computing hold promise for enhancing AI security in next-generation data science. Blockchain can provide a decentralized, immutable ledger for AI model updates and data transactions, ensuring transparency and integrity. Smart contracts can automate security policies and enforce compliance, reducing human error and tampering. While quantum computing poses a threat to current cryptographic methods, it also offers opportunities for quantum-resistant encryption algorithms. Quantum cryptography can provide ultra-secure communication channels, impervious to eavesdropping and interception.

AI security in next-generation data science is complex and dynamic, requiring a holistic and proactive approach. As AI permeates various data science domains, ensuring its security becomes increasingly vital. By implementing robust data governance frameworks, enhancing model resilience against adversarial attacks, and leveraging emerging technologies, we can safeguard AI systems and harness their full potential. The journey toward secure AI in data science is ongoing, but with continued innovation and vigilance, it promises significant benefits while mitigating associated risks.

This book constitutes the proceedings of the Third Southwest Data Science Conference (SDSC 2024), which took place in Waco, Texas, USA, on March 22–23, 2024.

All papers in this book were selected from 59 submissions that underwent a rigorous single-blind review process, with each paper being assigned to at least three reviewers. We accepted 15 full papers.

May 2024   Henry Han

Organization

General Chair

Henry Han   Baylor University, USA

Program Committee Chair

Henry Han   Baylor University, USA

Steering Committee

Henry Han   Baylor University, USA
Erich Baker   Belmont University, USA
Greg Hamerly   Baylor University, USA
Matthew Fendt   Baylor University, USA
Chen Zhao   Baylor University, USA
Candace Ditsch   Baylor University, USA
Patrick Hynan   Baylor University, USA

Program Committee

Hisham AI-Mubaid   University of Houston, USA
Mary Lauren Benton   Baylor University, USA
Erdogan Dogdu   Angelo State University, USA
Jeff Forest   Slippery Rock University, USA
Mark Ferguson   University of South Carolina, USA
Michael Gallaugher   Baylor University, USA
Chan Gu   Ball State University, USA
Keith Hubbard   Stephen F. Austin State University, USA
Xiuzhen Hu   Inner Mongolia University of Technology, China
David Kahle   Baylor University, USA
Haiquan Li   University of Arizona, USA
Dondong Li   South China University of Technology, China
David Li   Institute of Plant Physiology and Ecology, CAS,
                                  China

Zhen Li	Texas Woman's University, USA
Huiming Liu	Guangxi University, USA
Wenbin Liu	Guangzhou University, China
Michael Pokojovy	University of Texas, El Paso, USA
Eli Olinick	Southern Methodist University, USA
Guimin Qin	Xidian University, China
Greg Speegle	Baylor University, USA
Yang Sun	Park University, USA
James Stamey	Baylor University, USA
Ye Tian	Case Western Reserve University, USA
Stellar Tao	UT Health Center, USA
Tie Wei	Guangxi University, China
Jiacun Wang	Monmouth University, USA
Yi Wu	New York University, USA
Juanying Xie	Shaanxi Normal University, China
Honggang Zhang	University of Massachusetts, Boston, USA
Liang Zhao	University of Michigan, USA
Jeff Zhang	California State University, Northbridge, USA
Joe Zhou	Columbia University, USA

Additional Reviewers

Jie Ren
Xiang Fang
Yutao Fan

Contents

Wasserstein Graph Convolutional Network with Attention for Imbalanced scRNA-seq Data Knowledge Discovery

Jie Ren(ID) and Henry Han(✉)(ID)

The Laboratory of Data Science and Artificial Intelligence Innovation,
School of Engineering and Computer Science, Baylor University,
Waco, TX 76798, USA
henry_han@baylor.edu

Abstract. Discovering complex molecular patterns in imbalanced scRNA-seq data remains a challenge, despite numerous efforts from different perspectives. In this study, we propose a novel deep learning model: the Wasserstein Graph Convolutional Network with an attention mechanism (twGCN), designed for semi-supervised learning to address this challenge. The proposed model overcomes the weaknesses of traditional Graph Convolutional Networks by capturing more data intricacies and geometry. This is achieved by integrating a Wasserstein distance-based loss function optimization along with an attention mechanism. Unlike traditional scRNA-seq data preprocessing, we employ a robust scaling approach to normalize scRNA-seq data, which generally contains a large number of outliers. Our methods demonstrate significant advantages over peer methods in discovering single-cell patterns in benchmark data. More importantly, the proposed twGCN can handle both low-dimensional and high-dimensional scRNA-seq data obtained after feature selection. To our knowledge, this study will positively impact both deep learning and bioinformatics, inspiring future research.

Keywords: scRNA-seq · Wasserstein distance · data imbalance · Graph convolutional network · Attention mechanism · semi-supervised learning

1 Introduction

Single-cell omics technologies have dramatically transformed genomics research by unveiling unprecedented insights into cellular diversity and function across various conditions and species. As the volume of single-cell data escalates, efficiently integrating and analyzing this data becomes crucial for advancing our understanding of cellular behaviors. However, the inherent diversity in single-cell omics data—including variations across tissue types, species, and experimental conditions—complicates these efforts. Moreover, technical noise and inconsistencies between batches pose additional challenges, especially when using existing labeled datasets to interpret new ones.

H. Han (Ed.): SDSC 2024, CCIS 2158, pp. 1–16, 2024.
https://doi.org/10.1007/978-3-031-67871-4_1

High-dimensional single-cell omics technologies, such as scRNA-seq and scATAC-seq, present significant challenges in data integration and interpretation due to their complexity. Despite advancements in foundation models like scFoundation and scGPT [1, 2], which aim to analyze single-cell transcriptomic data, effectively integrating and interpreting this high-dimensional data remains difficult. The complexity of tumor ecosystems, as highlighted by Dezuani et al. [3], further complicates these efforts. This complexity is also evident in single-cell proteomics, where uncertainty estimation, data missingness, batch effects, and high noise hinder analysis. Li proposes scPROTEIN, a framework designed to address these issues in single-cell proteomics by estimating peptide quantification uncertainty and denoising data [4]. Similarly, Islam [5] highlights the challenges in single-cell genomics, where existing methods fail to consider underlying biological characteristics, compromising data analysis. To address this, an entropy-based cartography strategy, or genomap, has been developed to transform high-dimensional gene expression data into a spatial configuration that captures gene-gene interactions. Additionally, Luo et al. [6] emphasize the importance of preserving topological structure during dimensionality reduction in scRNA-seq data. Their single-cell graph autoencoder (scGAE) method effectively maintains this structure while enabling accurate cell clustering, trajectory inference, and data visualization.

However, important questions remain unanswered, such as how to handle data imbalance in single-cell RNA sequencing (scRNA-seq) data. Data imbalance in scRNA-seq arises from the unequal representation of cell types due to biological and technical factors, affecting downstream analyses like cell type identification and differential gene expression. Although different methods such as data augmentation, normalization, down-sampling, and weighting are proposed, they still cannot completely solve the scRNA-seq data imbalance issue, especially for large-scale data. scRNA-seq normalization methods are essential for minimizing biases and noise from low-input material, significantly impacting downstream analyses like classification and differential analysis [7]. However, current methods like SC3 (Single-Cell Consensus Clustering) and SCnorm have disadvantages such as the loss of subtle biological variation and potential overfitting, especially with low-abundance genes [8,9]. We also have even found that scRNA-seq normalization methods can introduce more biases when handling data imbalance.

To handle the challenge, this study proposes a novel deep learning model: the Wasserstein Graph Convolutional Network with an attention mechanism (twGCN), designed for semi-supervised learning to conduct knowledge discovery for imbalacned scRNA-seq data. Our contribution is threefold: Firstly, we view scRNA-seq data as a generic dataset with much outlier information and employ a Robust scaling way to conduct normalization rather than using current tailored scRNA-seq normalization methods [10,11]. Viewing scRNA-seq data as a generic dataset and using robust scaling for normalization enhances generalization, reduces biases, and effectively handles outliers, improving the reliability and accuracy of downstream analyses. This approach also ensures scalability and simplifies the preprocessing pipeline, making it more adaptable and reproducible.

Secondly, we innovate by incorporating the Wasserstein distance [12] into the loss function of the Graph Convolutional Network (GCN) model (scgcn) for handling scRNA-seq data via semi-supervised learning [13]. The Wasserstein distance, rooted in optimal transport theory, is well-suited for our needs as it captures the geometric characteristics of the data, enabling meaningful comparisons in high-dimensional spaces. Incorporating the Wasserstein distance into the loss function helps handle data imbalance by ensuring fair representation and focus on underrepresented cell types, reducing bias, and improving generalization. This approach also enhances the model's robustness to outliers and noise, leading to more reliable and accurate results

Finally, we integrate a transformer attention layer into the scgcn model significantly augmenting its capacity to discern and learn from the complex patterns in single-cell data [13]. By applying this attention mechanism, our model adapts to the intricate structures inherent in scRNA-seq data, potentially setting new benchmarks for prediction accuracy and the applicability of knowledge transfer techniques in single-cell omics. More importantly, adding an attention mechanism to the scGCN model helps handle data imbalance by dynamically focusing on underrepresented cell types, enhancing sensitivity to rare patterns, and mitigating bias. This leads to more balanced, accurate, and fair model performance in analyzing single-cell data.

This paper is structured as follows. We review related works in Sect. 2. Section 3 introduces the proposed twGCN model. Section 4 discusses the data and preprocessing. Section 5 covers the results of this study. Section 6 discusses the limitations of this work and presents our conclusion.

2 Related Work

Single-cell omics technologies, such as scRNA-seq and scATAC-seq, have transformed biomedical research at the cellular level. Despite the rapid growth of single-cell data, traditional label transfer methods like Seurat v3 [14], Conos [15], and CHETAH [16] struggle with dataset heterogeneity and primarily capture only straightforward similarities between cells. These methods often fail to account for complex topological relationships among cells, which are crucial for accurate data interpretation.

Ma presents heterogeneous graph transformers for single-cell biological network inference [17]. While DeepMAPS demonstrates superior performance in cell clustering and biological network construction, its computational complexity and high resource demands can pose significant challenges for widespread adoption. Furthermore, the dependency on high-quality scMulti-omics data may limit its effectiveness in cases with noisy or incomplete datasets.

Single-cell RNA sequencing provides high-throughput gene expression information to explore cellular heterogeneity at the individual cell level [18]. To address challenges related to dimensionality and the prevalence of dropout events, a deep graph learning method called scMGCA was developed. Despite its accuracy and effectiveness in cell segregation and batch effect correction, the complexity of scMGCA and its reliance on high-quality single-cell RNA sequencing data can limit its applicability in more challenging datasets with significant noise or missing values.

Graph Convolutional Networks (GCNs) have been introduced to overcome these limitations by utilizing the topological structures of cellular data to enhance performance. However, the application of GCNs [13] also faces challenges, particularly in handling the high variability and noise inherent in single-cell data, which can affect the stability and generalizability of the models.

Also, the advances in single-cell RNA sequencing (scRNA-seq) have spurred the development of clustering methods like SC3, which uses hierarchical clustering, and graph theory-based methods such as Seurat, that optimize cell cluster identification using mutual nearest neighbor (MNN) similarities [19]. Despite these improvements, many clustering approaches suffer from instability across different datasets and fail to adequately capture the complex topological relationships inherent in scRNA-seq data, limiting their effectiveness and adaptability.

3 Wasserstein Graph Convolutional Network with Attention (twGCN)

We introduce the proposed twGCN model in this section, which is an enhancement of the single-cell Graph Convolutional Network (scGCN) from [13]. The scGCN approach uses single-cell data to infer cellular-level knowledge in query datasets through semi-supervised learning. It constructs a sparse and hybrid graph of cell mappings using mutual nearest neighbors, enabling shared information propagation between reference and query data. This allows scGCN to project both datasets onto the same latent space, accurately predicting and learning cell labels in the query data from the reference data.

3.1 twGCN Architecture

The architecture of our model, illustrated in Fig. 1, is specifically designed to process and analyze heterogeneous datasets in single-cell genomics. The diagram illustrates the multi-layered structure, starting with input graph construction that integrates reference and query datasets. It progresses through graph convolutional layers, including a transformer attention layer, and culminates in hidden and output layers that finalize cell label predictions. Solid lines represent intra-dataset connections, while dashed lines indicate inter-dataset connections, facilitating comprehensive data integration and analysis.

3.2 Wasserstein Loss Function

In a GCN, we have a graph $G = (V, E)$, where V is the set of nodes and E is the set of edges. Each node $v \in V$ has a feature vector $\mathbf{x}_v$ and potentially a label $\mathbf{y}_v$.

The goal of the GCN is to learn a function $f : \mathbb{R}^{|V| \times d} \to \mathbb{R}^{|V| \times c}$ that maps the input node features to output labels, where d is the dimensionality of the input features and c is the number of classes.

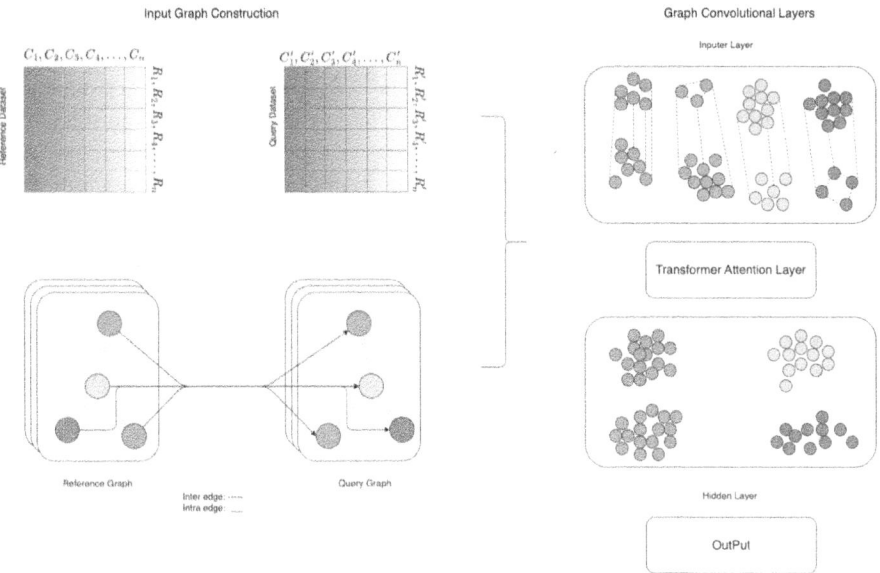

Fig. 1. Architecture of the twGCN model - This diagram illustrates the multi-layered architecture of the model, starting with the input graph construction that integrates reference and query datasets. It details the progression through graph convolutional layers, including a transformer attention layer, and culminates in the hidden and output layers that finalize cell label predictions. Solid lines represent intra-dataset connections, while dashed lines indicate inter-dataset connections, facilitating comprehensive data integration and analysis.

The cross-entropy loss for a single node v is defined as:

$$CE(\mathbf{y}_v, \hat{\mathbf{y}}_v) = -\sum_{j=1}^{c} y_{v,j} \log(\hat{y}_{v,j}) \tag{1}$$

where:

- $\mathbf{y}_v$ is the one-hot encoded true label vector for node v.
- $\hat{\mathbf{y}}_v$ is the predicted probability vector for node v.
- $y_{v,j}$ is the true label indicator for class j (1 if node v belongs to class j, 0 otherwise).
- $\hat{y}_{v,j}$ is the predicted probability for class j.

The overall cross-entropy loss for the GCN is the average of the losses over all labeled nodes:

$$\mathcal{L}_{CE} = \frac{1}{|V_L|} \sum_{v \in V_L} CE(\mathbf{y}_v, \hat{\mathbf{y}}_v) \tag{2}$$

where V_L is the set of labeled nodes.

A key issue with the classic cross-entropy loss function is its reliance on the Euclidean distance, which can produce biased results for high-dimensional data such as scRNA-seq data [20]. Additionally, the cross-entropy loss function struggles to handle imbalanced datasets in classification tasks, leading to problematic decisions based on these predictions.

We use the Wasserstein distance as our loss function due to its inherent geometric properties, which offer several advantages. Firstly, it is defined on a metric space, highlighting the spatial relationships and distances between points. This characteristic ensures a more accurate representation of the underlying data structure. Additionally, the Wasserstein distance is highly sensitive to the shape and geometry of distributions. This sensitivity makes it an exceptional tool for comparing complex, high-dimensional data distributions, as it captures subtle differences that other distance measures might overlook.

For high-dimensional data, such as scRNA-seq, the Euclidean distance used in cross-entropy can be misleading. The Wasserstein distance considers the overall distribution structure, providing a more accurate measure. In a high-dimensional space $\mathbb{R}^n$, the Wasserstein distance accounts for the spatial arrangement of data points:

$$W_1(P, Q) = \inf_{\gamma \in \Gamma(P,Q)} \int_{\mathbb{R}^n \times \mathbb{R}^n} \|x - y\| \, d\gamma(x, y) \tag{3}$$

The Wasserstein loss is sensitive to the shape of distributions. In contrast, cross-entropy focuses on pointwise probabilities, potentially ignoring structural similarities between distributions:

$$L_{\text{CE}} = -\sum_i y_i \log(\hat{y}_i) \tag{4}$$

Here, y_i and $\hat{y}_i$ are the true and predicted probabilities, respectively. This formulation does not consider the geometry or overall distribution shape.

3.3   Graph Construction in GCN

The model begins with the input graph constructed from reference and query datasets, visualized through two matrices: the reference dataset matrix $\mathbf{M}_r$ and the query dataset matrix $\mathbf{M}_q$. Each matrix's rows and columns represent cells and their respective gene expressions, which are mapped to construct the hybrid graph. This graph consists of two distinct parts: the reference graph and the query graph. The graph convolution construction is defined as:

$$H^{(l+1)} = \sigma \left(\tilde{D}^{-1/2} \tilde{A} \tilde{D}^{-1/2} H^{(l)} W^{(l)} \right) \tag{5}$$

where:

- $H^{(l)}$ is the matrix of node features at layer l.
- $\tilde{A} = A + I$ is the adjacency matrix with added self-loops.

- $\tilde{D}$ is the degree matrix of $\tilde{A}$, with $\tilde{D}_{ii} = \sum_j \tilde{A}_{ij}$.
- $W^{(l)}$ is the layer-specific trainable weight matrix.
- σ is an activation function, such as ReLU.

The connections within these graphs are inter-edges, which connect data within each reference and query dataset, and intra-edges, which connect data between the reference and query datasets.

3.4   Graph Convolutional Networks with Attention Mechanism

Following the input graph construction, the model includes graph convolutional layers. The transformer attention layer applies a multi-head attention mechanism to focus on the most relevant features across the graph:

$$
\text{Attention}(Q, K, V) = \frac{e^{\frac{QK^T}{\sqrt{d_k}}}}{\sum_j e^{\frac{QK_j^T}{\sqrt{d_k}}}} V \tag{6}
$$

where:

- Q (query), K (key), and V (value) are linear transformations of the input.
- d_k is the dimension of the key vectors.

Each head computes attention independently, and the results are concatenated and linearly transformed to form the final output. These layers are responsible for processing the graph data to capture and encode the complex relationships and patterns. The layers include:

- **Input Layer:** Graph data that initializes features and begins the transformation process.
- **Transformer Attention Layer:** Applies transformer-based multi-head attention mechanisms [21] to focus on and prioritize the most relevant features across the graph. Each head in the multi-head attention mechanism computes self-attention separately, focusing on different parts of the input.
- **Hidden Layers:** Further processes the information, refining feature representations and preparing the final output. This layer aggregates the enhanced features, ensuring that similar cells with the same labels cluster together, promoting accurate label predictions for the query data.
- **Output Layer:** Produces the final predictions, which are the categorized labels for the cells in the query dataset.

Overall, this architecture leverages the strengths of graph convolutional networks and transformer attention mechanisms to address the challenges posed by the high dimensionality and variability of single-cell genomic data. This sophisticated approach ensures that the model is not only capable of handling complex data structures but also effective in predicting and learning from them. The overall model aggregates these layers, resulting in the following sequential operations:

$$H^{(1)} = \sigma(\tilde{D}^{-1/2}\tilde{A}\tilde{D}^{-1/2}XW^{(0)}) \qquad (7)$$

$$H^{(2)} = \text{Attention}(H^{(1)}W_Q, H^{(1)}W_K, H^{(1)}W_V) \qquad (8)$$

$$H^{(3)} = \sigma(\tilde{D}^{-1/2}\tilde{A}\tilde{D}^{-1/2}H^{(2)}W^{(1)}) \qquad (9)$$

The final output layer produces the predictions:

$$\text{Output} = \frac{\exp(H^{(3)}W^{(2)})}{\sum_j \exp(H^{(3)}W_j^{(2)})} \qquad (10)$$

Overall, this architecture leverages the strengths of graph convolutional networks and transformer attention mechanisms to address the challenges posed by the high dimensionality and variability of single-cell genomic data. This sophisticated approach ensures that the model is not only capable of handling complex data structures but also effective in predicting and learning from them.

4   Data and Preprocessing

4.1   Data Imbalance in scRNA-seq data

To demonstrate our model's performance, we selected two benchmark imbalanced scRNA-seq dataset: SRP073767 and Pollen datasets.

The SRP073767 dataset contains single-cell RNA sequencing data from human samples, generated using the 10x Genomics Chromium platform. It includes indexed sequencing libraries from 29 germline and cancer samples, sequenced with advanced technologies such as the NextSeq 500 and Illumina HiSeq 2500.

The Pollen dataset [22] comprises 249 cells spanning 11 different populations, including various neural and blood cells. Its primary purpose is to evaluate the effectiveness of low-coverage single-cell RNA sequencing in distinguishing distinct cell populations. It features a diverse array of cell types, such as skin cells, pluripotent stem cells, blood cells, and neural cells. Within this dataset, Class8 has the highest representation, whereas Class5 is among the least represented.

The imbalance in both datasets poses challenges for machine learning models. In the SRP073767 dataset, majority classes ('beta' and 'alpha') may overshadow minority classes (e.g., 'macrophage'), leading to biased models that perform poorly on underrepresented classes. Similarly, in the Pollen dataset, the dominance of certain classes can affect the model's ability to accurately predict and analyze less represented cell types. Without proper balancing techniques, these imbalances could undermine the model's overall performance and reliability.

As illustrated in Fig. 2, the SRP073767 dataset predominantly consists of beta cells, while the Pollen dataset has a higher count of Class8. In contrast, macrophages and Class5 are the minority classes in the SRP073767 and Pollen datasets, respectively, underscoring the data imbalance in both.

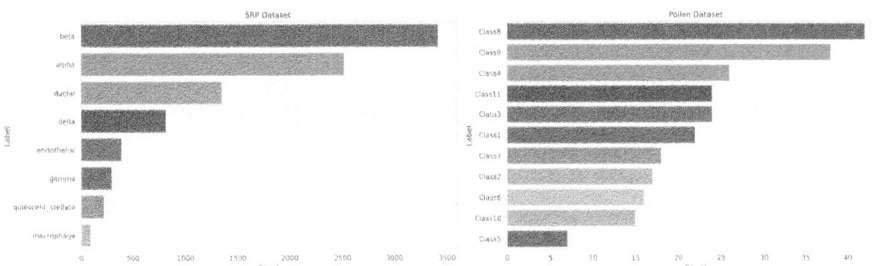

Fig. 2. Distribution of cell types in the SRP073767 dataset and the distribution of classes in the Pollen dataset. The SRP073767 dataset predominantly consists of beta cells, while the Pollen dataset has a higher count of Class8. In contrast, macrophages and Class5 are the minority classes in the SRP073767 and Pollen datasets, respectively, highlighting the data imbalance in both datasets.

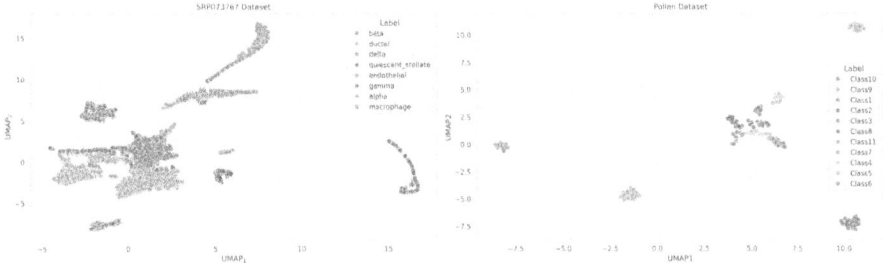

Fig. 3. UMAP plots visualizing the clustering of cell types in the SRP073767 dataset and the clustering of classes in the Pollen dataset. The SRP073767 dataset shows distinct clusters for beta, ductal, delta, quiescent stellate, endothelial, gamma, alpha, and macrophage cells, indicating clear separations among these cell types. Similarly, the Pollen dataset reveals distinct clusters for Class 1 through Class 11, with separations between classes.

4.2   Feature Representation

The UMAP plots illustrate Fig. 3 the clustering patterns within the SRP073767 and Pollen datasets. In the SRP073767 dataset, distinct clusters for cell types such as beta, ductal, and delta suggest clear separations and potentially unique biological functions. Similarly, the Pollen dataset displays well-separated clusters for its classes, emphasizing the distinct characteristics of each class. Using UMAP for manifold learning [23], we visualized the feature representation of the SRP073767 and Pollen datasets. This approach effectively captures the diversity and imbalance within these datasets, providing insights into their underlying biological or categorical distinctions.

4.3 Data Preprocessing

Our data preprocessing approach follows the methodology described in [13]. The input dataset is divided into two primary categories. The first category, reference data, includes samples with known labels. This data is used to train the models and serves as a benchmark for evaluating annotation performance. The second category, query data, consists of unlabeled samples that require annotation based on the predictions of the trained model.

To identify the most impactful features in the reference data, we employ two main strategies: Analysis of Variance (ANOVA) to compare gene expression across multiple classes, followed by the selection of the top 2000 genes with the most significant adjusted P-values for further analysis.

- **ANOVA Application:** We conduct an Analysis of Variance (ANOVA) on the reference data to compare gene expression means across multiple classes. This method identifies genes with significant expression differences, which is crucial for understanding cellular processes.
- **Gene Selection:** Post-ANOVA, the top 2000 genes with the most statistically significant adjusted P-values are selected. This ensures that our model incorporates genes that are both statistically relevant and likely to contribute to meaningful insights.

After these steps, we analyze data sparsity, as summarized in Table 1. In high-dimensional spaces, data points tend to be spread out sparsely, making it difficult to identify patterns and relationships due to the low density of data points in any given region. This sparsity can lead to challenges in deep learning, as the models may struggle to learn effectively from the sparse data and make accurated result [15,19]. Sparsity is quantified as the proportion of zero entries relative to the total number of entries in the dataset. This is an inherent characteristic of high-dimensional, sparse single-cell RNA sequencing (scRNA-seq) data, primarily caused by dropout events during sequencing. Higher sparsity means a greater number of zero values, which can influence subsequent analyses. Typically, the sparsity of an scRNA-seq dataset falls between 25% and 85% in the absence of imputation processing.

Table 1. Summary of Datasets

Dataset	No. of Cells	No. of Genes	No. of Classes	Sparsity (%)
Pollen	249	2000	11	47.04
SRP073767	9000	2000	8	80.65

Table 1 illustrates the varying degrees of sparsity across different datasets. For instance, the Pollen dataset, with 249 cells and 2000 genes, exhibits a sparsity of 47.04%, indicating a moderate level of zero entries. In contrast, the SRP073767 dataset, containing 9000 cells and 2000 genes, shows a higher sparsity of 80.65%. This significant difference underscores the importance of addressing data sparsity in scRNA-seq analysis, as it can impact the performance of various computational methods and the interpretation of biological results.

4.4 Normalization Method

Single-cell RNA sequencing (scRNA-seq) data is characterized by high dimensionality, sparsity due to drop-out events, and inherent noise and variability arising from the low amount of starting material and biological differences among cells. Due to these factors, scRNA-seq data often contains numerous outliers. To address this, we applied the robust scaler to our dataset for its resilience to outliers and ability to transform data to a consistent range.

Normalization methods such as scran and DESeq2 are commonly used in scRNA-seq data analysis. However, these methods have their disadvantages. Scran, while effective at normalizing cell-specific biases, can be computationally intensive and may not handle outliers well [24]. DESeq2, which assumes a negative binomial distribution, may not be suitable for datasets with high dropout rates and extreme variability [25]. Each scRNA-seq normalization method can introduce different biases, making it crucial to understand these biases to select the most appropriate method for a given analysis.

Unlike standard scaling, which is influenced by outliers, the robust scaler uses the median and interquartile range (IQR) for scaling. This approach makes it more robust to outliers, preventing distortion of the data scale. The robust scaling method works by subtracting the median and then dividing by the IQR. Mathematically, for a feature vector $\mathbf{x}$:

$$\mathbf{x}_{\text{scaled}} = \frac{\mathbf{x} - \text{median}(\mathbf{x})}{\text{IQR}(\mathbf{x})} \tag{11}$$

where $\text{median}(\mathbf{x})$ is the median of the feature vector $\mathbf{x}$, and $\text{IQR}(\mathbf{x})$ is the interquartile range, calculated as the difference between the 75th and 25th percentiles of $\mathbf{x}$. This method ensures that the data is scaled in a way that is robust against the influence of outliers, making it ideal for datasets with significant outliers.

5 Result

Table 2 underscores the superior accuracy of Our Model at 97.87% under TensorFlow 2, significantly outpacing the baseline model, scGCN (90% accuracy under TensorFlow v1.0), as well as other methods such as Seurat v3 (87% accuracy). In stark contrast, CHETAH exhibits the lowest accuracy at 62%, reflecting the wide performance variance among the evaluated models. The comparative results presented in the table are derived from [13], with the scGCN model utilizing TensorFlow v1.0 at that time.

Table 2. Comparison of accuracy of different models

Method	Accuracy
scGCN	90%
Seurat v3	87%
Conos	83%
scmap	82%
CHETAH	62%
Our Model	97.87%

5.1   D-index

We use accuracy and D-index to measure our model and baseline model performance. Accuracy is a metric used in machine learning to measure the ratio of correct predictions made by a model out of all predictions made. Although accuracy is widely used, it has some limitations when used with imbalanced datasets, In imbalanced datasets, models can display misleadingly high accuracy scores by predicting the majority class while neglecting the minority class. However high accuracy does not necessarily indicate effective performance in minority classes. Additionally, accuracy alone fails to provide insights into the types of errors being made, such as false positives and false negatives, which are crucial for a comprehensive understanding of model performance.

In contrast to traditional metrics like accuracy and the F1 score, which may be biased or misleading in imbalanced datasets, the D-index [26] can provide a more comprehensive assessment of machine learning model behaviors. This index is particularly effective in distinguishing states of imbalance, with values ranging within (0,2]. In multiclass case, the D-index measures model performance by calculating the expected value of local index values from all classes. This provides a comprehensive metric that effectively captures the model's ability to handle diverse class distributions, especially for minority classes:

$$d = \frac{1}{k} \sum_{i=1}^{k} \left(\log_2(1 + \alpha_i) + \log_2 \left(1 + \frac{s_i + p_i}{2} \right) \right) \qquad (12)$$

where α_i is the accuracy, s_i is specificity, and p_i is precision for different i in class k.

5.2   Ablation Study

In this ablation study, we aim to analyze the performance of various methods applied to different datasets, with a particular focus on the scGCN model introduced by [13]. We use this model as our baseline to evaluate enhancements through modifications and additional components.

Table 3 presents a comprehensive comparison of performance metrics across four methods: scGCN, GCN with attention mechanism , GCN with Wasserstein loss, and twGCN with both attetion mechansim and Wasserstein loss. The experiments are conducted on two distinct datasets, SRP073767 and Pollen, using TensorFlow v2 for implementation.

For the SRP073767 dataset, the baseline scGCN model achieves an accuracy of 0.93, a D-Index of 1.91, a precision of 0.79, a recall of 0.82, and an F1-score of 0.80. Introducing attention mechanisms to the GCN (GCN + Attention) improves the accuracy to 0.95, the D-Index to 1.95, precision to 0.94, recall to 0.90, and F1-score to 0.92, indicating significant performance gains. Further enhancement with the Wasserstein distance metric (GCN + Wasserstein) yields an accuracy of 0.97, a D-Index of 1.98, precision of 0.97, recall of 0.98, and an F1-score of 0.97. The twGCN model, which combines these improvements, achieves the highest performance with an accuracy of 0.98, a D-Index of 1.99, precision of 0.96, recall of 0.98, and an F1-score of 0.97.

The Pollen dataset exhibits a different performance trend. The baseline scGCN model shows significantly lower metrics, with an accuracy of 0.25, a D-Index of 1.38, precision of 0.04, recall of 0.17, and an F1-score of 0.07. The GCN + Attention model significantly boosts the accuracy to 0.88, D-Index to 1.91, precision to 0.86, recall to 0.86, and F1-score to 0.93. The GCN + Wasserstein model further improves these metrics, achieving an accuracy of 0.92, a D-Index of 1.93, precision of 0.95, recall of 0.90, and an F1-score of 0.90. The twGCN model outperforms the others with an accuracy of 0.94, a D-Index of 1.95, precision of 0.97, recall of 0.92, and an F1-score of 0.93.

These results indicate that the incorporation of attention mechanisms and the Wasserstein distance metric significantly enhance the performance of the GCN models across both datasets. The twGCN model consistently demonstrates the highest performance metrics, suggesting that the combined approach is the most effective.

Table 3. Performance Metrics Comparison Across Different Methods and Datasets

DataSet	Method	Accuracy	D-Index	Precision	Recall	F1-score
SRP073767	scGCN	0.93	1.91	0.79	0.82	0.80
SRP073767	GCN + Attention	0.95	1.95	0.94	0.90	0.92
SRP073767	GCN + Wasserstein	0.97	1.98	0.97	0.98	0.97
SRP073767	twGCN	0.98	1.99	0.96	0.98	0.97
Pollen	scGCN	0.25	1.38	0.04	0.17	0.07
Pollen	GCN + Attention	0.88	1.91	0.86	0.86	0.93
Pollen	GCN + Wasserstein	0.92	1.93	0.95	0.90	0.90
Pollen	twGCN	0.94	1.95	0.97	0.92	0.93

6   Discussion and Conclusion

The Tables 2 and 3 demonstrate that twGCN consistently outperforms other methods across different datasets, achieving the highest accuracy, D-Index, and F1-score. Specifically, twGCN achieves an accuracy of 97.87% on the SRP073767 dataset and 94% on the Pollen dataset.

Despite these advancements, our twGCNs with attention mechanisms and Wasserstein distance-based loss functions have potential weaknesses. The integration of an attention mechanism within GCNs increases the computational load, as each node's attention calculation involves other nodes, potentially leading to quadratic growth in computation with the number of nodes. Additionally, the Wasserstein distance is computationally intensive, especially in high-dimensional spaces, as calculating this distance requires solving an optimization problem. Furthermore, these models are prone to overfitting, particularly when dealing with complex and high-dimensional datasets, which can limit their generalizability and performance on unseen data.

To address these challenges, we propose the following solutions:

– **Scalable Attention Mechanisms:** Use sparse attention mechanisms or hierarchical attention to reduce the computational burden. These techniques approximate full attention by focusing only on a subset of important nodes or using multi-scale approaches.
– **Approximate Wasserstein Distance:** Employ approximations or variants of the Wasserstein distance that are computationally cheaper, such as the Sinkhorn distance, which allows for more efficient computation while retaining the benefits of the Wasserstein metric.

In this work, we explored alternative normalization methods, specifically employing Robust Scaler Normalization to handle outliers. We incorporated the Wasserstein distance into the loss function to capture geometric data characteristics more effectively than the Euclidean distance. Additionally, we enhanced the Graph Convolutional Network (GCN) with a transformer attention layer, enabling the model to learn complex patterns in single-cell RNA sequencing (scRNA-seq) data. These advancements improve prediction accuracy and enhance the applicability of knowledge transfer techniques in single-cell omics.

Acknowledgement. This work is partially supported by NASA Grant 80NSSC22 K1015, NSF 2229138, and McCollum endowed chair startup fund.

References

1. Hao, M., Gong, J., Zeng, X., et al.: Large-scale foundation model on single-cell transcriptomics. Nat. Methods 1–11 (2024)
2. Cui, H., Wang, C., Maan, H., et al.: scGPT: toward building a foundation model for single-cell multi-omics using generative AI. Nat. Methods (2024)
3. De Zuani, M., Xue, H., Park, J.S., et al.: Single-cell and spatial transcriptomics analysis of non-small cell lung cancer. Nat. Commun. **15**, 4388 (2024)

4. Li, W., Yang, F., Wang, F., et al.: scPROTEIN: a versatile deep graph contrastive learning framework for single-cell proteomics embedding. Nat. Methods **21**, 623–634 (2024)
5. Islam, M.T., Xing, L.: Cartography of genomic interactions enables deep analysis of single-cell expression data. Nat. Commun. **14**, 679 (2023)
6. Luo, Z., Chang, X., Zhang, Z., et al.: A topology-preserving dimensionality reduction method for single-cell RNA-seq data using graph autoencoder. Sci. Rep. **11**, 20028 (2021)
7. Lytal, N., Ran, D., An, L.: Normalization methods on single-cell RNA-seq data: an empirical survey. Front. Genet. **11**, 41 (2020)
8. Kiselev, V.Y., et al.: SC3: consensus clustering of single-cell RNA-seq data. Nat. Methods **14**(5), 483–486 (2017)
9. Bacher, R., Chu, L.F., Leng, N., et al.: SCnorm: robust normalization of single-cell RNA-seq data. Nat. Methods **14**, 584–586 (2017)
10. Hafemeister, C., Satija, R.: Normalization and variance stabilization of single-cell RNA-seq data using regularized negative binomial regression. Genome Biol. **20**(1), 296 (2019)
11. Sharma, V.: A study on data scaling methods for machine learning. Int. J. Glob. Acad. Sci. Res. **1**(1), 31–42 (2022)
12. Arjovsky, M., Chintala, S., Bottou, L.: Wasserstein generative adversarial networks. In: International Conference on Machine Learning, pp. 214–223 (2017)
13. Song, Q., Jing, S., Zhang, W.: scGCN is a graph convolutional networks algorithm for knowledge transfer in single cell omics. Nat. Commun. **12**(1), 3826 (2021)
14. Stuart, T., et al.: Comprehensive integration of single-cell data. Cell **177**(7), 1888–1902 (2019)
15. Barkas, N., et al.: Joint analysis of heterogeneous single-cell RNA-seq dataset collections. Nat. Methods **16**, 695 (2019)
16. de Kanter, J.K., Lijnzaad, P., Candelli, T., Margaritis, T., Holstege, F.C.P.: CHETAH: a selective, hierarchical cell type identification method for single-cell RNA sequencing. Nucleic Acids Res. **47**(16), e95 (2019)
17. Ma, A., et al.: Single-cell biological network inference using a heterogeneous graph transformer. Nat. Commun. **14**(1), 964 (2023)
18. Yu, Z., Su, Y., Lu, Y., et al.: Topological identification and interpretation for single-cell gene regulation elucidation across multiple platforms using scMGCA. Nat. Commun. **14**, 400 (2023)
19. Guan, J., Li, R.-Y., Wang, J.: Grace: a graph-based cluster ensemble approach for single-cell RNA-seq data clustering. IEEE Access **8**, 166730–166741 (2020)
20. Aggarwal, C.C., Hinneburg, A., Keim, D.A.: On the surprising behavior of distance metrics in high dimensional space. In: Database Theory-ICDT 2001: 8th International Conference London, UK, 4–6 January 2001 Proceedings 8, LNCS, pp. 420–434. Springer, Heidelberg (2001). https://doi.org/10.1007/3-540-44503-X_27
21. Vaswani, A., et al.: Attention is all you need. In: Guyon, I., Von Luxburg, U., Bengio, S., Wallach, H., Fergus, R., Vishwanathan, S., Garnett, R. (eds.) Advances in Neural Information Processing Systems, vol. 30. Curran Associates Inc., New York (2017)
22. Pollen, A.A., et al.: Low-coverage single-cell mRNA sequencing reveals cellular heterogeneity and activated signaling pathways in developing cerebral cortex. Nat. Biotechnol. **32**(10), 1053–1058 (2014)
23. Han, H., Li, W., Wang, J., Qin, G., Qin, X.: Enhance explainability of manifold learning. Neurocomputing **500**, 877–895 (2022)

24. Lun, A.T.L., McCarthy, D.J., Marioni, J.C.: A step-by-step workflow for low-level analysis of single-cell RNA-seq data with bioconductor. F1000Research **5**, 2122 (2016)
25. Love, M.I., Huber, W., Anders, S.: Moderated estimation of fold change and dispersion for RNA-seq data with deseq2. Genome Biol. **15**(12), 550 (2014)
26. Han, H., Yi, W., Wang, J., Han, A.: Interpretable machine learning assessment. Neurocomputing **561**, 126891 (2023)

Toward a Unified Cybersecurity Knowledge Graph: Leveraging Ontologies and Open Data Sources

Adam Boyer[✉], Erdogan Dogdu, Roya Choupani, Jason S. Watson,
Diego Sanchez, and Alexander Ametu

Department of Computer Science, Angelo State University, San Angelo, USA
{aboyer,edogdu,rchoupani,jason.watson,dsanchez35,aametu}@angelo.edu
http://www.cs.angelo.edu

Abstract. The cybersecurity field is very heterogeneous with the ever-growing digital cyberspace and the increasing volume of big data produced in the field. It is therefore difficult to overcome many of the security challenges with manual solutions. Automation and intelligent cybersecurity solutions are both promising tools to overcome common security challenges, however, they require robust data and knowledge management. Today, Knowledge graphs (KG) are widely used in many intelligent system solutions, including the cybersecurity field. However, the efforts to build KGs in many cybersecurity areas are narrowly focused, and therefore it is difficult to unify and integrate these otherwise very useful solutions. Earlier efforts in unifying the cybersecurity knowledge using common ontologies have not generalized their approach to provide a unified solution for cybersecurity challenges. Here, we attempt provide a renewed approach to building a Unified Cybersecurity Knowledge Graph (UCKG), using the Unified Cybersecurity Ontology (UCO), that integrates structured and unstructured open data sources in an automated fashion. With this paper, we hope to pave the way toward a Unified Cybersecurity Knowledge Graph (UCKG) and its utilization in intelligent cybersecurity solutions.

Keywords: Cybersecurity ontology · cybersecurity knowledge graph · intelligent cybersecurity · Unified Cybersecurity Ontology · Unified Cybersecurity Knowledge Graph

1 Introduction

There has been an explosion of data and computational power in recent years, and almost every domain now benefits from big data, data science, machine learning, and AI in many ways. Cybersecurity is no exception and it benefits tremendously from big data and AI [14]. Big data and data-driven AI transform the cybersecurity domain tremendously. We see more data-driven and intelligent solutions brought to the cybersecurity field in all areas, such as security

© The Author(s), under exclusive license to Springer Nature Switzerland AG 2024
H. Han (Ed.): SDSC 2024, CCIS 2158, pp. 17–33, 2024.
https://doi.org/10.1007/978-3-031-67871-4_2

assessment, threat discovery, and all sorts of other intelligent security operations [2–10].

The current cybersecurity landscape is full of data and knowledge sources, pouring from many resources, from government agencies to industrial bodies. The heterogeneous nature of this data space underlines the need for a unified knowledge resource that can be used to provide comprehensive and intelligent cybersecurity solutions. Previous research has focused on solutions in specific areas such as malware detection, intrusion detection, etc. While efforts to solve specific issues plaguing the cyberseurity domain is useful, a unified solution is needed to provide comprehensive cybersecurity solutions.

The number and breadth of cybersecurity threats is ever-changing. Up-to-date information on the latest threats is needed to maintain the usefulness of any cybersecurity solution. With distributed sources of information, it is difficult to correlate resources in real time without a unified knowledge source. There needs to be comprehensive, integrated, and up-to-date source of information available that connects disparate resources into a single pane of glass.

Ontologies and knowledge graphs are important tools in unifying the distributed and disparate cybersecurity information sources and data. There has been efforts toward to this goal earlier [1,3,9], but these early efforts failed to complete their objectives and abandoned the initiatives early. Now, with the increasing interest in the utilization of knowledge graphs in all types of intelligent solutions, there is a chance to start this over and provide a comprehensive and working data integration solution for cybersecurity. This time the possibility of success is higher, since the field has matured in the last few years in a number of fronts. First, KGs are being utilized and has been shown to provide success in many machine learning and intelligent automated solutions. So, there is an opportunity to utilize KGs in the cybersecurity field, there has been many efforts aleready, we just have to show a working unified solution. Second, the cybersecurity field has grown tremendously with the growth and availability of data. Now is the time to harvest. There is need for a comprehensive solution that integrates big data using KG representation and reasoning. Third, we have reason to believe that the earlier efforts in unified cybersecurity ontologies and vocabularies did not get wide adoption due to the immaturity of the field and the up and coming age of AI. Now, with big data and sophisticated ontology solutions, we believe there will be many intelligent applications taking advantage of KGs for many different scenarios, from risk assessment to cyber threat intelligence modeling, cyber readiness and defense, and more.

In this paper, we address the need for a unified cybersecurity knowledge graph and present a method for building one using the Unified Cybersecurity Ontology (UCO) and its enhancements. We envision integrating scattered data sources such as Common Vulnerabilities Enumeration (CVE), Common Weakness Enumeration (CWE), Common Attack Pattern Enumeration and Classification (CAPEC), Common Platform Enumeration (CPE), Common Vulnerability Scoring System (CVSS), and After Action Reports (AARs). This is a more comprehensive list of resources than earlier efforts. Our UCKG is also extensible, new data sources can be added and integrated using a common ontology. We also present a vision for how the ontology and accompanying KG can be utilized in intelligent cybersecurity operations.

In Sect. 2 we present the previous work in this area. We review the ontology we are using and the integration approach in Sect. 3 and then present the methodology we are using to build the UCKG in Sect. 4. Section 5 reviews the list of structured and unstructured data sources we are using to integrate and build the UCKG. Section 6 presents an approach for visualizing and exploring the UCKG which we plan to integrate into a comprehensive application with this project. There could be many intelligent applications benefiting from and using the UCKG, and we present a glimpse of those in Sect. 7. We finally conclude and point to future work in Sect. 8.

2   Related Work

Cybersecurity knowledge graphs (CKG) have been getting a lot of attention recently, due to the cybersecurity field using and benefiting from big data and intelligent solutions[13,15–17]. CKGs are utilized to represent knowledge obtained from cybersecurity field. They use a graph-model to represent concepts, entities, and artifacts as nodes, and the relationships between them as edges. CKGs are used to represent complex cybersecurity knowledge and data in a simple and unified framework for easy management, access, and maintenance. CKGs also allow us to develop sophisticated AI-based intelligent cybersecurity solutions.

Zhao et al. surveyed the area of CKG construction [8]. They reviewed studies constructing (1) general CKGs that can be used in many types of applications, and (2) those CKGs that are constructed for specific application areas, such as intrusion detection, malware detection, risk assessment, etc. They also point to the gap between abstract CKG information and actual log data, as well as the scarcity of abnormal data within.

There are many uses of KGs in cybersecurity and many studies have developed and utilized CKGs for different purposes. For example, Piplai et al. [11] presented a system for extracting information from "After Action Reports" (AAR), building a cybersecurity knowledge graph (CKG) and fused information coming from multiple AAR reports and resources, therefore presenting an integrated view of threat intelligence.

STIX [3] is an earlier effort from MITRE corporation in standardizing cybersecurity vocabulary. However, STIX is based on XML, a legacy standard for building semantic knowledge bases or knowledge graphs. STIX also incorporated vocabularies from other standards, therefore it was valuable effort in the path towards a unified ontology. The UCO also incorporates and maps concepts and entities from STIX [9].

STUCCO[1] is yet another effort in providing a comprehensive CKG. It is an organized ontology schema collecting information from many structured and

---

[1] https://stucco.github.io/.

unstructured data sources [1]. GitHub repository[2] of the project shows little to no activity since its release.

The Unified Cybersecurity Ontology[3] (UCO) [9] is one of the latest ontologies to support information integration in cybersecurity, incorporating heterogeneous data and knowledge schemata from disparate cybersecurity systems and commonly used cybersecurity standards for information sharing and exchange. The UCO can be used to represent temporal (time) information and therefore promising reasoning capabilities. It can be used in information extraction from textual reports, blogs, and similar data sources, and thus speed up the process of information exchange and developing countermeasures against cyber attacks. The latest released version the UCO 1.5[4] includes more than 500 classes and 900 properties. According to a recent study by Bolton et al [12] the most commonly used ontology for building CKGs is UCO, so it is adopted well in the community.

SEPSES [20] is a recent CKG effort to develop a comprehensive dataset with data collected from many open data sources, such as CVE, CWE, CAPEC, CPE, using a standard ontology like UCO. A bottom-up approach is followed in order to build the SEPSES CKG using National U.S. Vulnerability Database (NVD) and a set of online security resources. They also demonstrated the effectiveness of SEPSES by two use case studies in vulnerability assessment and intrusion detection. However, there is no activity in their project[5] since its inception, version 1.1.0.

It is clear that there is a strong need for a unified CKG, both for providing a unified ontology or vocabulary to integrate disconnected cybersecurity knowledge sources, and also for collecting the artifacts, actual data representing the cybersecurity-related domain information and data. Here we attempt to do that with both structured and unstructured open data sources and a standard ontology, like UCO.

Liu et al. [2] pointed out that although there has been a large number of work in building CKG for different purposes, there is limited work on how to utilize CKG in applications to solve real-world problems.

Sarker et al. presented the most recent developments in AI-driven cybersecurity [10]. They review the most popular studies in the area of machine learning, deep learning, and natural language processing toward intelligent cybersecurity decision making, analysis, and management. They also discuss research opportunities in knowledge representation and rule-based expert system modeling.

We plan to provide a number of use cases as future work to showcase the utilization of the UCKG in intelligent cybersecurity solutions.

---

[2] https://github.com/stucco.

[3] https://github.com/Ebiquity/Unified-Cybersecurity-Ontology.

[4] https://github.com/Ebiquity/Unified-Cybersecurity-Ontology/blob/master/uco_1_5.owl.

[5] https://github.com/sepses/cyber-kg-converter.

3   Cybersecurity Ontology, Integration, and Project UCKG

Ontologies are an essential components of CKGs. Ontologies allow subject matter experts and modelers to create an agreed-upon structure and vocabulary for sharing information in different domains [18]. They allow machines to better understand the information contained in a knowledge graph and to make inferences. An example of a widely adopted ontology is the Financial Industry Business Ontology[6], or FIBO, which was created as a standard of terms and relationships used in financial institutions.

The Unified Cybersecurity Ontology (UCO) is the most complete and comprehensive at the time of this writing [9]. The UCO ontology includes the following classes for categorizing cybersecurity entities: Means, Consequences, Attack, Attacker, AttackPattern, Exploit, Exploit Target, and Indicator. These classes were derived from cybersecurity standards such as CVE, CWE, CAPEC, CPE, and CVSS. The UCO was meant to be a semantic extension of the STIX project, which was very comprehensive, but did not allow reasoning and used the legacy XML format.

Because the UCO is based on current cybersecurity standard metrics, we believe that it is a adequate ontology for integrating the current publicly available cybersecurity information into a "Unified Cybersecurity Knowledge Graph" (UCKG). The goal of our project, UCKG, is to use the UCO ontology to integrate CVE, CWE, CAPEC, CPE, and CVSS into a knowledge graph in an automated way. The UCO's vocabulary will be used to dynamically map and transform data into a unified, connected, integrated cybersecurity knowledge graph, which will be a reference for future intelligent cybersecurity applications.

4   Methodology

We take inspiration for the architecture of UCKG and our knowledge graph creation from the methods described by Kiesling et al [20] for their SEPSES knowledge graph. Unlike the SEPSES project, we follow the Unified Cybersecurity Ontology proposed earlier. We also leveraged the Neo4j plugin Neosemantics[7] to help import our RDF data into the Neo4j graph. Figure 1 shows a high-level overview of our UCKG creation and automated update process.

4.1   Automated Data Acquisition

Multiple scripts are written to download the initial data from CVE, CWE, CAPEC, CPE, and CVSS. Once the initial construction of the knowledge graph is complete, a task scheduler is used to poll the respective data sources twice a day for updates.

---

[6] https://edmcouncil.org/frameworks/industry-models/fibo/.
[7] https://neo4j.com/labs/neosemantics/.

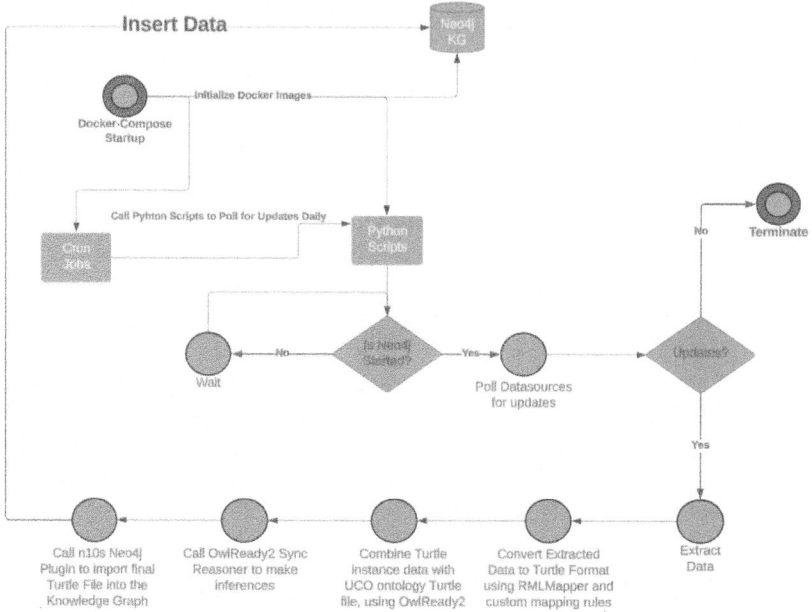

Fig. 1. Knowledge Graph Creation Flow.

4.2 Transform Data to RDF

The data received from the data sources comes in the form of JSON and XML. We transform the data into RDF, using Turtle format. The RML mapping language[8] is the language used for mapping structured data to RDF format. We then create these mappings and process the data acquired using a simple RML engine RMLStreamer[9].

4.3 Combination of UCO Ontology with Instance Data

Our project leverages the Python library OwlReady2[10] to combine our generated cybersecurity instance data with the UCO ontology. Once the instance data and the ontology are in loaded into one Turtle file, we call the sync reasoner, provided by OwlReady2, to infer any edges between instance nodes.

---

[8] https://rml.io/docs/rml/introduction/.

[9] https://github.com/RMLio/RMLStreamer.

[10] https://owlready2.readthedocs.io/en/v0.42/.

4.4   Importing Data into Neo4j

As previously stated, Neo4j is used as our graph database or triple store. Once the ontology and instance data are combined into a single Turtle file, we utilize the Neosemantics plugin to import our data into Neo4j. We load the data in batches to allow for logging and monitoring of the process.

4.5   Current State of the UCKG

Currently, the UCKG source code can be downloaded from our public GitHub repository[11]. Instructions for compiling and running the UCKG are presented as a README file. We will soon add an initial Turtle format snapshot of the UCKG to the repository. Figure 2 is a Cypher query showing how related CPE, CWE, CPE, CAPEC, and MITRE Att&CK can be extracted from the UCKG, using a given CVE ID. Figure 3 is the resulting graph from the Cypher query. Table 1 and Fig. 4 display the current totals for nodes and relationships, along with the Cypher query used to extract the data, respectively.

```
 1 MATCH
 2 (exploit:UcoExploitTarget)
 3     -[:UCOHASVULNERABILITY]→
 4         (vul:UcoVulnerability {uri:"http://purl.org/cyber/uco#VULN-CVE-2008-1901"}),
 5 (exploit:UcoExploitTarget)
 6     -[:UCOHASWEAKNESS]→
 7         (cwe:UcoCWE {uri: "http://purl.org/cyber/uco#CWE-59"}),
 8 (capec:UcoCAPEC)
 9     -[:UCOHASRELATEDWEAKNESS]→
10        (cwe:UcoCWE),
11 (capec:UcoCAPEC)
12     -[:UCOHASTAXONOMYMAPPING]→
13        (attack:UcoMITREATTACK),
14 (vul:UcoVulnerability)
15     -[:UCOHASCVE_ID]→
16        (cve:UcoCVE),
17 (cve:UcoCVE)
18     -[:HASCPE]→
19        (cpe:UcoCPE)
20 RETURN vul, exploit, cwe, cve, cpe, capec, attack
```

Fig. 2. Cypher Query For Given CVE

4.6   Future Enhancements

One of the goals of our project is to provide a unified data source for cybersecurity information. We understand that installing a local version of the UCKG which needs to be initialized and then kept running to poll for updates is not

---

[11] https://github.com/edogdu/UCKG.

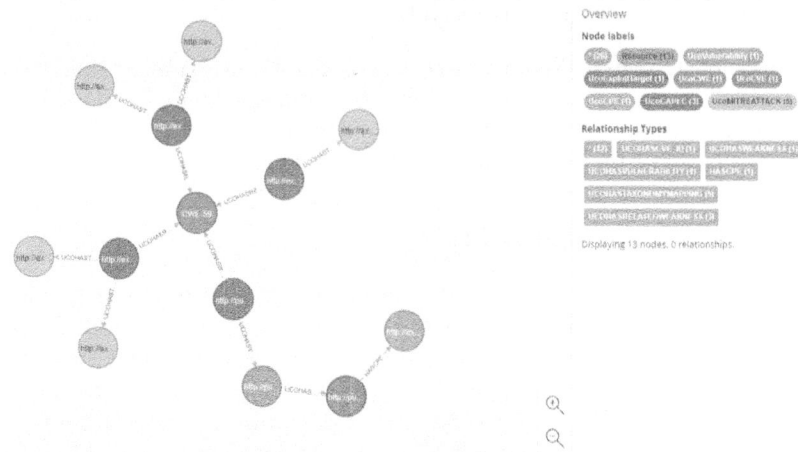

Fig. 3. Resulting Graph of CVE Query

optimal. To make setting up the UCKG less cumbersome, we plan to separate out the initialization and update scripts from the local copy. We plan to use a CI/CD workflow, possibly Git Actions[12], in combination with a topic and subscriber model to push new updates to client UCKGs. Figure 5 shows a high-level overview of our UCKG creation and update flow, after implementing a topic/subscriber model.

5   Cybersecurity Data Resources

Many of the publicly available cyber security data sources are linked to each other by properties or object IDs. Figure 6 depicts how these different data sources are linked together. We briefly describe each data source below.

5.1   Common Vulnerabilities Enumeration (CVE)

The Common Vulnerabilities List, or CVE, is a publicly available catalog of known cybersecurity vulnerabilities. The list first became publicly available in September 1999. Analysts and researchers can access the CVE List via a direct download on the official CVE website[13], or through the CVE API hosted on the NIST website[14]. Currently, there is a total of 215095 CVE records available. A CVE record will be assigned a CVE-ID by a CVE Numbering Authority

---

[12] https://docs.github.com/en/actions.

[13] https://www.cve.org/.

[14] https://nvd.nist.gov/developers/products.

Table 1. UCKG Totals

Description	Total
Total Nodes	638,031
Total Relationships	2,057,745
CVE Nodes	245,747
CWE Nodes	959
CPE Nodes	129,560
CAPEC Nodes	177
MITRE ATT&CK Nodes	830
MITRE D3FEND Nodes	193

(CNA). CVE records contain a description of the vulnerability, references to better understand the vulnerability, and the name of the CNA. CVE records are crucial for security experts and vendors to recognize and mitigate security vulnerabilities (Table 2).

Table 2. Important CVE Fields

Field	Example
CVE-ID	CVE-2023-5257
Description	A vulnerability was found in ...
References	https://vuldb.com/?id.240866
Assigning CNA	VulDB
Date Record Created	20230929

5.2   Common Weakness Enumeration (CWE)

The Common Weakness Enumeration, or CWE, is a list of common software and hardware weakness types. There are currently over 600 categories in the CWE list. CWEs include useful information such as a description, relationships to other CWEs, applicable platforms, and many more The CWE list is community-developed, and was first released in 2006. It can be accessed via a direct download from the CWE website[15]. Efforts are currently being taken to standardize the CWE into a JSON-based REST API[16], but as of the time of this writing, it is not complete. The CWE is a useful tool for researchers and analysts to recognize common flaws that may affect software or hardware (Table 3).

---

[15] https://cwe.mitre.org/index.html.
[16] https://cwe.mitre.org/documents/CWE-CAPEC_Rest_API_Working_Group.pdf.

```
 1  MATCH (n)
 2  RETURN 'Total Nodes' AS Description, COUNT(n) AS Total
 3      UNION ALL
 4  MATCH ()-[r]-()
 5  RETURN 'Total Relationships' AS Description, COUNT(r) AS Total
 6      UNION ALL
 7  MATCH (cve:UcoCVE)
 8  return ('CVE Nodes') as Description, COUNT(cve) AS Total
 9      UNION ALL
10  MATCH (cwe:UcoCWE)
11  return ('CWE Nodes') as Description, COUNT(cwe) AS Total
12      UNION ALL
13  MATCH (cpe:UcoCPE)
14  return ('CPE Nodes') as Description, COUNT(cpe) AS Total
15      UNION ALL
16  MATCH (capec:UcoCAPEC)
17  return ('CAPEC Nodes') as Description, COUNT(capec) AS Total
18      UNION ALL
19  MATCH (attack:UcoMITREATTACK)
20  return ('MITRE Att&ck Nodes') as Description, COUNT(attack) AS Total
21      UNION ALL
22  MATCH (defend:UcoMITRED3FEND)
23  return ('MITRE D3FEND Nodes') as Description, COUNT(defend) AS Total
24
```

Fig. 4. Cypher Query for UCKG Metrics

Table 3. Important CWE Fields

Field	Example
Weakness ID	787
Weakness Name	Out-of-bounds Write
Description	The product writes data past the end of the intended buffer
Relationships	ChildOf CWE-119
Applicable Platforms	C (Often Prevalent)

5.3   Common Attack Pattern Enumeration and Classification (CAPEC)

The Common Attack Pattern Enumeration and Classification, or CAPEC, was initially established by the U.S. Department of Homeland as an effort to identify and share known attack patterns in the cybersecurity community. CAPEC was initially released in 2007 and allows for public contributions to its knowledge base. There are currently 559 total attack patterns listed by CAPEC. CAPEC records list a brief description of the attack pattern, severity of the attack, possible mitigation methods, related CWEs, and more. The CAPEC list can be accessed via a direct download from the CAPEC website[17]. Efforts are currently

---

[17] https://capec.mitre.org/index.html.

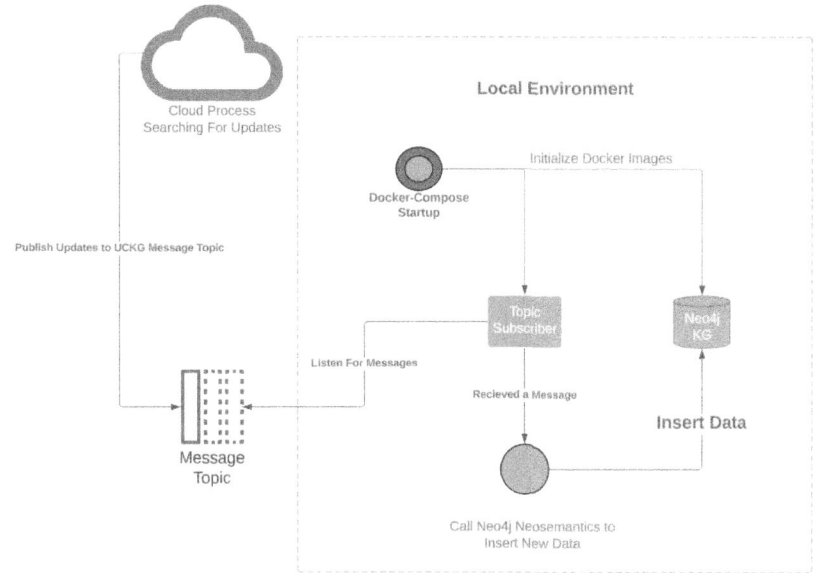

Fig. 5. Topic/Subscriber UCKG Flow

being taken to standardize the CAPEC into a JSON-based REST API (See foot-note 11), but as of the time of this writing, it is not complete. The CAPEC list is needed by analysts and researchers to be able to recognize the signs of a possible attack (Table 4).

Table 4. Important CAPEC Fields

Field	Example
Category ID	156
Category Name	Engage in Deceptive Interactions
Summary	Attack patterns within this category focus on ...
Membership	MemberOf ID 1000
Submission Date	2014-06-23

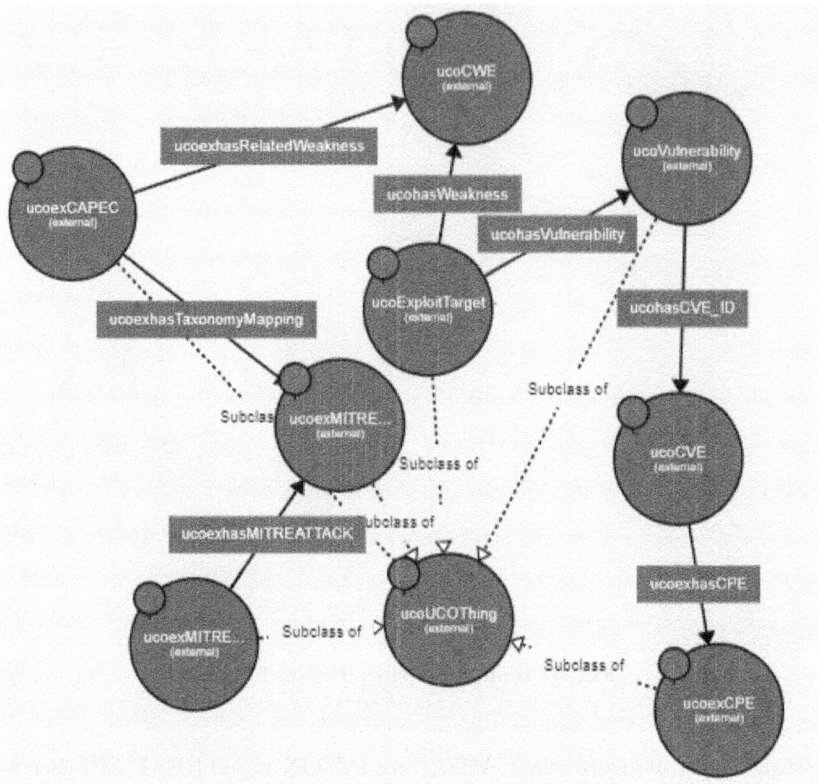

Fig. 6. Data source connected diagram.

5.4   Common Platform Enumeration (CPE)

The Common Platform Enumeration, or CPE, is a publicly available dictionary of known systems, software, and packages. Naming of entities in the dictionary is based on the Uniform Resource Identifier, URI, format. Some important information contained in the CPE are CPE version, part or type (i.e. applications, hardware, or operating systems), vendor, and product. The CPE dictionary is continually being updated, with over 16,000 additions made in just September of 2023. Analysts and researchers can access the CPE dictionary via a direct download[18] or through the CPE API hosted on the NIST website(See footnote 2). The CPE dictionary is essential for cybersecurity experts to have a structured way of identifying entities (Table 5).

5.5   Common Vulnerability Scoring System (CVSS)

The Common Vulnerability Scoring System, or CVSS, is a standardized framework used to describe the severity of software vulnerabilities. The CVSS is owned

---

[18] https://nvd.nist.gov/products/cpe.

Table 5. Important CPE Fields

Field	Example
CPE Name	cpe:2.3:a:3com:3cserver:-:*:*:*:*:*:*:*
CPE ID	FFADC2E2-6888-490E-98CB-6D86D2E893A6
Last Modified	2011-01-12T14:35:43.867
Title	3Com 3CServer

and maintained by the non-profit organization FIRST.Org Inc[19]. The CVSS score is based on three metrics: Base, Temporal, and Environmental. The base score can be between 0 and 10. The score corresponds to an associated severity score: 0 is None, 0.1–3.9 is Low, 4.0–6.9 is Medium, 7.0–8.9 is High, and 9.0–10.0 is Critical. The National Institute of Standards and Technology, or NIST, has published a CVSS calculator[20] for CVEs on their website. CVSS scores can help cybersecurity experts prioritize which vulnerabilities to address first (Table 6).

Table 6. Important CVSS Fields

Field	Example
Vector String	CVSS:3.1/AV:N/AC:H/PR:L/
Base Score	3.1 (Low)
Temporal Score	3.0 (Low)
Environment Score	3.0 (Low)

5.6   After Action Reports (AARs)

Piplai et al. [11] describe After Action Reports as being a good resource for extracting cyber attack information. AARs are created as unstructured text documents, and traditionally cyber analysts would need to read the documents to extract useful information manually. Figure 7 depicts an example AAR. Piplai et al [11] describe a system for extracting useful information from the reports using natural language processing (NLP) techniques and populating a knowledge graph with this information. Having these AARs in a structured format would allow researchers and analysts to find common attack patterns or trends automatically using intelligent tools.

---

[19] https://www.first.org/cvss/.
[20] https://nvd.nist.gov/vuln-metrics/cvss/v3-calculator.

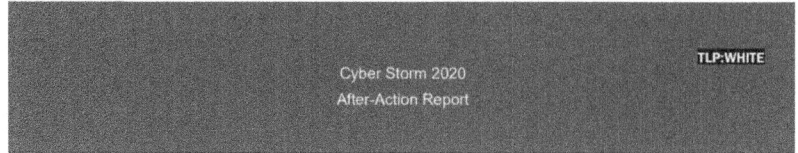

EXECUTIVE SUMMARY

Exercise Background

On August 10-14, 2020, the Cybersecurity and Infrastructure Security Agency (CISA) conducted Cyber Storm 2020 (CS 2020), the seventh iteration of the national capstone cyber exercise that brings together the public and private sectors to simulate response to a cyber crisis impacting the Nation's critical infrastructure. Cyber Storm exercises are part of CISA's ongoing efforts to assess and strengthen cyber preparedness and examine incident response processes. The exercise findings contribute to safeguarding the Nation's security and cyber infrastructure by identifying ways to strengthen coordinated incident response along the whole-of-Nation approach outlined in the *National Cyber Incident Response Plan* (NCIRP). CISA sponsors the exercise series to improve capabilities of the cyber incident response community, encourage the advancement of public-private partnerships within the critical infrastructure sectors, and strengthen the relationship between the Federal Government and its government partners at the state, local, and international levels.

Exercise Goal & Objectives

> **Goal:** Strengthen cybersecurity preparedness and response capabilities by exercising policies, processes, and procedures for identifying and responding to a multi-sector cyberattack targeting critical infrastructure.

Exercise Objectives:

- Examine the implementation and effectiveness of national cybersecurity plans and policies;

- Strengthen and enhance information sharing and coordination mechanisms used across the cyber ecosystem during a cyber incident;

Fig. 7. An excerpt from an After Action Report [19].

6 Knowledge Graph Ontology Visualization

We provide a visualization[21] of the UCKG to help users dynamically explore, interact, and better understand the ontology and the UCKG. WebVOWL[22] is an open-source graphical user interface that provides this exact functionality. Classes are represented as labeled nodes in a graph connected by relationships. By clicking on various nodes and relationships, users will be shown detailed information about the selection including name, type, comments, and who defined the entity. Figure 8 shows a screenshot from WebVOWL.

7 Intelligent Cybersecurity Applications Utilizing Knowledge Graphs

Various applications can be constructed that utilize an extensive knowledge graph containing platforms, weaknesses, and vulnerabilities. One major area is the study of attack flow, how an attack develops and early recognition of the signs of an attack occurrence. Another possible application of an extensive cyber

---

[21] https://service.tib.eu/webvowl/#iri=https://raw.githubusercontent.com/edogdu/ UCKG/main/data/UCKG_Snapshots/uco_extended.ttl

[22] http://vowl.visualdataweb.org/webvowl.html.

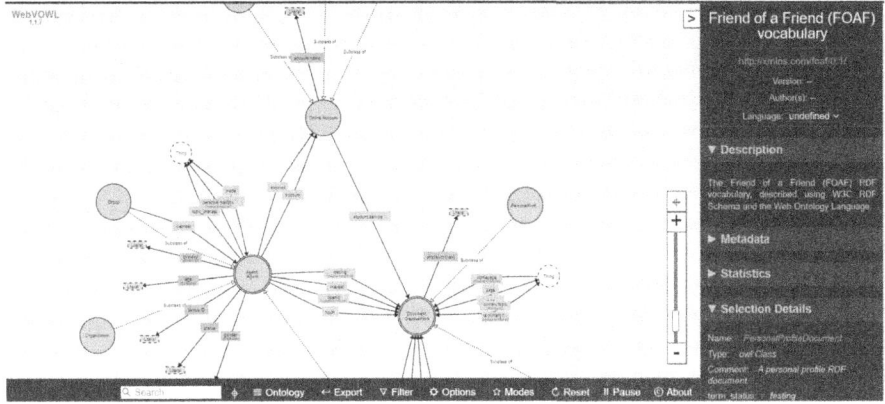

Fig. 8. Screenshot of WebVOWL user interface.

threat knowledge graph is the discovery of new vulnerabilities and threats. This includes the personalization of threat discovery specific to organizations with unique systems and cyber infrastructure. Log-based analysis as compared to an extensive knowledge base is another section of future study related to threat detection. Predicting Zero-Day attacks based on similar characteristics and patterns of current known cyber attacks could greatly benefit the general community of security professionals as well as offer an individual approach to preemptively protect company assets and data. Finally, an attack pattern knowledge graph can be used alongside intrusion detection systems to monitor traffic and predict live attacks that will happen or are currently in progress [2]. We intend to develop several use cases as future work to show the effectiveness of UCKG in real-world intelligent applications.

8 Conclusion and Future Work

In this paper, we presented UCKG, Unified Cybersecurity Knowledge Graph, a comprehensive KG towards collecting, integrating, unifying cybersecurity knowledge resources. The aim of UCKG is to provide a single point of reference for future intelligent cybersecurity applications. Our prototype application currently integrates data from open cybersecurity data resources such as CPE, CWE, CAPEC, CPE, and CVSS. UCO is the unifying vocabulary and ontology to unify integrated data sources and to build UCKG.

We plan to add AAR reports using NLP solutions and possibly Large Language Models (LLM). There is also high interest in mapping platform-specific logs, such as Microsoft SQL Server[23] logs in UCKG.

---

[23] https://www.microsoft.com/en-us/sql-server.

8.1 Future Integration with Large Language Models and After Action Reports

Large Language Models (LLMs) aid cybersecurity operations by efficiently processing vast and diverse knowledge sources. They automate tasks such as threat report analysis, contextualization of threat intelligence for specific organizations, and generation of organization-specific threat responses, thereby streamlining processes and improving overall efficiency in threat detection, mitigation, and response.

Liu and Zhan [21] introduce an efficient method for creating knowledge graphs from Cyber Threat Intelligence (CTI) reports by utilizing a Large Language Model (LLM) like ChatGPT. The approach automates the extraction of attack-related entities and their relationships, outperforming existing methods like AttacKG and REBEL in terms of performance while requiring fewer manual interventions and computational resources, showing its viability in low-resource scenarios within the cyber threat intelligence domain.

Mitra et al. [22] introduces LocalIntel, an automated system that contextualizes threat reports for specific organizations by retrieving global threat data and integrating it with local knowledge. LocalIntel operates in three phases: retrieving global threat intelligence, accessing relevant local knowledge, and combining both sources through a generator to generate organization-specific threat responses.

In this project, our aim is to further integrate a module that utilizes Large Language Models (LLMs), such as ChatGPT, to analyze textual cybersecurity data, particularly "After Action Reports" (AAR), extracting relevant information and associating it with the Universal Cyber Knowledge Graph (UCKG).

References

1. Iannacone, M., et al.: Developing an ontology for cyber security knowledge graphs. In: Proceedings of the 10th Annual Cyber and Information Security Research Conference, pp. 1–4 (2015)
2. Liu, K., Wang, F., Ding, Z., Liang, S., Yu, Z., Zhou, Y.: Recent progress of using knowledge graph for cybersecurity. Electronics 11(15), 2287 (2022)
3. Barnum, S.: Standardizing cyber threat intelligence information with the structured threat information expression (STIX). Mitre Corp. 11, 1–22 (2012)
4. Guo, Y., Liu, Z., Huang, C., Wang, N., Min, H., Guo, W., Liu, J.: A framework for threat intelligence extraction and fusion. Comput. Secur. 132, 103371 (2023)
5. Keshavarzi, M., Ghaffary, H.R.: An ontology-driven framework for knowledge representation of digital extortion attacks. Comput. Hum. Behav. 139, 107520 (2023)
6. Qi, Y., et al.: Cybersecurity knowledge graph enabled attack chain detection for cyber-physical systems. Comput. Electr. Eng. 108, 108660 (2023)
7. Bryniarska, A., Pokuta, W.: Ontology-based knowledge representation in the IoT cybersecurity system. Semantic Web J. (2022)
8. Zhao, X., Jiang, R., Han, Y., Li, A., Peng, Z.: A survey on cybersecurity knowledge graph construction. Comput. Secur. 103524 (2023)
9. Syed, Z., Padia, A., Finin, T., Mathews, L., Joshi, A.: UCO: a unified cybersecurity ontology. UMBC Student Collection (2016)

10. Sarker, I.H., Furhad, M.H., Nowrozy, R.: AI-driven cybersecurity: an overview, security intelligence modeling and research directions. SN Comput. Sci. **2**, 1–18 (2021)
11. Piplai, A., Mittal, S., Joshi, A., Finin, T., Holt, J., Zak, R.: Creating cybersecurity knowledge graphs from malware after action reports. IEEE Access **8**, 211691–211703 (2020). https://ieeexplore.ieee.org/stamp/stamp.jsp?arnumber=9264152
12. Bolton, J., Elluri, L., Joshi, K.: An overview of cybersecurity knowledge graphs mapped to the MITRE ATT&CK framework domains. UMBC Center for Accelerated Real-Time Analysis (2023)
13. Sikos, L.F.: Cybersecurity knowledge graphs. Knowl. Inf. Syst. 1–21 (2023)
14. Piplai, A.: Knowledge Graphs and Reinforcement Learning: A Hybrid Approach for Cybersecurity Problems (Doctoral dissertation, University of Maryland, Baltimore County) (2023)
15. Andrew, Y., Lim, C., Budiarto, E.: Knowledge graphs for cybersecurity: a framework for honeypot data analysis. In: 2023 IEEE International Conference on Cryptography, Informatics, and Cybersecurity (ICoCICs), pp. 275–280. IEEE (2023)
16. Piplai, A., Kotal, A., Mohseni, S., Gaur, M., Mittal, S., Joshi, A.: Knowledge-enhanced Neuro-Symbolic AI for Cybersecurity and Privacy (2023). arXiv preprint arXiv:2308.02031
17. Wang, P., Liu, J., Hou, D., Zhou, S.: A cybersecurity knowledge graph completion method based on ensemble learning and adversarial training. Appl. Sci. **12**(24), 12947 (2022)
18. Allemang, D., Hendler, J.: Semantic Web for the Working Ontologist: Effective Modeling in RDFS and OWL. Elsevier, Amsterdam (2011)
19. Cybersecurity and Infrastructure Security Agency. Cyber Storm 2020 After-Action Report (2020). https://fsscc.org/wp-content/uploads/2021/02/Cyber_Storm-2020_After-Action-Report_01052021_Final.pdf
20. Kiesling, E., Ekelhart, A., Kurniawan, K., Ekaputra, F.: The SEPSES knowledge graph: an integrated resource for cybersecurity. In: International Semantic Web Conference, pp. 198–214. Springer, Cham (2019). https://doi.org/10.1007/978-3-030-30796-7_13
21. Liu, J., Zhan, J.: Constructing knowledge graph from cyber threat intelligence using large language model. In: 2023 IEEE International Conference on Big Data (BigData), pp. 516–521. IEEE (2023)
22. Mitra, S., et al.: LOCALINTEL: generating organizational threat intelligence from global and local cyber knowledge. arXiv preprint arXiv:2401.10036 (2024)

How Does Normalization Impact Clustering?

Ashley Han[1](✉) and Hongrui Du[2]

[1] Skyline High School, Ann Arbor, MI 48105, USA
ashleythehan@gmail.com
[2] Shanghai World Foreign Language Academy, Shanghai 200233, China

Abstract. Normalization is becoming increasingly important with the surge in data complexity and volume in AI. Despite its significance, normalization often does not receive adequate attention. In this study, we evaluate various normalization methods on benchmark datasets across different clustering algorithms to understand their impacts on clustering. Contrary to the general assumption that density-based clustering methods are more robust to the effects of normalization than partitioning clustering methods, our results show that partitioning methods demonstrate greater efficiency and robustness under different normalization methods compared to density-based methods. Density-based methods may provide less reliable results, even when parameters are fine-tuned for well-structured data. Furthermore, density-based methods can produce significantly skewed clustering results for imbalanced data, regardless of the normalization methods used. However, partitioning methods can still achieve solid performance in handling imbalanced data. Our study sheds light on the development of more customized normalization methods for different clustering scenarios.

Keywords: Normalization · Clustering · Machine learning · Data imbalance

1 Introduction

In the rapidly evolving field of Artificial Intelligence (AI), the complexity and volume of data have surged dramatically. As datasets grow in size and diversity, the need for effective preprocessing techniques becomes increasingly critical. One such technique is normalization, a process that adjusts the scale of data features to a common range. This is essential for ensuring that no single feature disproportionately influences the outcome of AI models, particularly in different machine learning such as clustering applications [1].

Normalization transforms data to bring all features into comparable ranges, facilitating more accurate and efficient computations. This preprocessing step is crucial in scenarios where data features have different units or magnitudes, as it ensures that each feature contributes equally to the downstream analysis,

© The Author(s), under exclusive license to Springer Nature Switzerland AG 2024
H. Han (Ed.): SDSC 2024, CCIS 2158, pp. 34–47, 2024.
https://doi.org/10.1007/978-3-031-67871-4_3

whether it be a machine learning procedure or another type of analysis. Without normalization, machine learning models and data analysis algorithms might produce biased or even incorrect results, leading to poor model performance.

It remains unknown less understood that the impact of normalization on clustering algorithms. Some previous works investigate the impacts of normalization methods in the classification for different machine learning models [1,2]. However, it is rarely investigated topic for clustering, especially more and more new normalization and clustering models have been proposed in various fields [3,4]. Unlike classification, where the objective is to map inputs to predefined labels, clustering aims to discover the inherent structure in the data without prior knowledge of the labels. This makes the choice of normalization method critical, as it can significantly alter the clustering results.

Clustering methods can be grouped into partitioning methods and density-based methods. Partitioning clustering methods divide data into different clusters according to certain criteria, typically by optimizing an objective function such as minimizing the sum of squared distances within clusters. Partitioning methods can be further divided into flat partitioning and hierarchical partitioning methods, depending on whether the clusters have a hierarchical structure. The former includes methods such as K-means, GMM (Gaussian Mixture Models), Self-Organizing Maps (SOM), Affinity Propagation, and Spectral Clustering; the latter includes methods such as Bisecting K-means and hierarchical clustering [6].

Density-based clustering methods divide data into clusters based on regions with varying densities of data points. These methods identify clusters as areas of high point density separated by regions of low density, allowing them to effectively detect clusters of arbitrary shapes and handle noise in the data, The density-based methods include DBSCAN ((Density-Based Spatial Clustering of Applications with Noise), HDBSCAN (Hierarchical Density-Based Spatial Clustering of Applications with Noise), OPTICS (Ordering Points To Identify the Clustering Structure) and their variants [7].

There is a general assumption, but without serious verification, that density-based methods are more robust to the effects of normalization than partitioning methods in achieving good performance. This is mainly because of their focus on local density rather than global distances. For example, density-based clustering algorithms, such as DBSCAN and HDBSCAN, define clusters based on the density of data points in a local neighborhood. This local approach makes the clustering outcome less influenced by the overall scale of the data after normalization and more by the relative distances and density of points in small regions. On the other hand, partitioning methods, such as K-means, rely heavily on global distance metrics to assign points to clusters. The global distance metrics can be significantly impacted by normalization methods.

In this study, we evaluate the impacts of normalization on benchmark datasets using various partitioning and density-based clustering algorithms. Contrary to the general assumption, we have found that density-based clustering methods, such as DBSCAN, HDBSCAN, and OPTICS, are more sensitive to normalization methods than partitioning clustering methods, such as K-means and Bisecting K-means (BiKmeans). Partitioning methods demonstrate greater efficiency and robustness under different normalization methods compared to density-based methods. In contrast, density-based methods may provide less reliable results even when parameters are fine-tuned for well-structured data. Furthermore, density-based methods can produce significantly skewed clustering results for imbalanced data regardless of the normalization method used. However, partitioning methods can still achieve solid performance in handling imbalanced data.

Our findings unveil the impacts of normalization methods on clustering. They will guide the selection of clustering models, especially for imbalanced data, which is essential in finance, medicine, and cybersecurity fields. Understanding that partitioning methods are more robust to normalization biases helps ensure more reliable and consistent clustering outcomes. By challenging the general assumption about the robustness of density-based methods, the study contributes to a deeper understanding of clustering algorithms and their dependencies on normalization techniques. It sheds light on the development of more customized normalization methods for different clustering scenarios. These findings can also serve as a foundation for future research aimed at developing new clustering models and their matched normalization algorithms for imbalanced data.

2   Method

We employ widely used normalization and clustering methods across benchmark datasets and use three clustering metrics to evaluate clustering performance. In this section, we briefly introduce normalization methods, clustering methods, and clustering evaluation metrics.

2.1   Normalization Methods

We employ four normalization methods in this study: standard scaling, min-max scaling, max-abs scaling, and robust scaling. Standard scaling is a model-driven normalization method where data is assumed to follow a normal distribution. In contrast, the other three methods are purely data-driven and do not assume any specific data distribution.

Standard scaling transforms each feature to have a mean of zero and a standard deviation of one by using the following formula,

$$x' = \frac{x - \mu}{\sigma} \tag{1}$$

where x is the incoming feature, μ and σ are the mean and standard deviation of the feature. This method assumes data follows a Gaussian distribution and may not be suitable for non-Gaussian data. It is also sensitive to outliers and can produce distorted scaling because both μ and σ are sensitive to outliers.

Minmax scaling scales the data to a fixed range, typically [0, 1]. This method is useful when the distribution of the data is not Gaussian and when the range of values needs to be bounded. Min-max scaling is calculated as follows:

$$x' = \frac{x - \min(x)}{\max(x) - \min(x)} \tag{2}$$

This maps the minimum and maximum values of each feature to the specified range, preserving relative relationships between values. It works well for data with similar ranges but different units, such as in images and finance. However, min-max scaling can be highly sensitive to outliers, as extreme values can disproportionately affect the scaling, compressing the majority of the data into a narrow range.

Maxabs scaling scales each feature by its maximum absolute value: $x' = \frac{x}{\max |x|}$. This method maps all values within the range $[-1, 1]$. Max-abs scaling works well for well-structured data such as those data centered around zero and when it is important to preserve existing structure of the data, such as in text data represented by word frequencies.

Robust scaling works for datasets with outliers or skewed distributions. It transforms the data based on the median and the interquartile range (IQR), using the formula:

$$x' = \frac{x - \text{median}(x)}{\text{IQR}(x)} \tag{3}$$

where the interquartile range (IQR) is the difference between the 75th percentile (Q_3) and the 25th percentile (Q_1)

Robust scaling is advantageous for features with varying distributions, as it reduces the impact of extreme values and ensures balanced normalization. Using the median and IQR, it preserves the central tendency and spread of the data, making it reliable for datasets with irregular distributions. However, it can not work well for data without outliers, especially for homogeneous data. It can also miss important information beyond the IQR.

2.2 Partitioning and Density-Based Clustering Methods

We employ two groups of clustering methods in this study: partitioning and density-based methods. The partitioning methods include K-means and Bisecting K-means, with K-means being a flat partitioning method and Bisecting K-means being a hierarchical partitioning method. The density-based methods include DBSCAN (Density-Based Spatial Clustering of Applications with Noise), HDBSCAN (Hierarchical Density-Based Spatial Clustering of Applications with Noise), and OPTICS (Ordering Points To Identify the Clustering Structure).

K-means partitions input data into k clusters, where each observation belongs to the cluster with the nearest mean. The algorithm minimizes the within-cluster sum of squares (WCSS), defined as:

$$WCSS = \sum_{i=1}^{k} \sum_{x \in C_i} \|x - \mu_i\|^2 \tag{4}$$

where C_i is the set of points in cluster i, and μ_i is the mean of points in C_i. K-means iteratively updates cluster centroids and reassigns points to the nearest centroid until convergence.

While very intuitive and easy to implement, K-means assumes the number of clusters is known, and it is sensitive to initialization and noise, and it fails on non-convex data and does not handle high-dimensional data well.

Bisecting K-means (BiKmeans) integrates hierarchical clustering and K-means. Starting with all data points in a single cluster, it recursively splits the cluster into two sub-clusters using K-means, and then selects the cluster with the largest variance (or within-cluster sum of squares) to split again. This process continues until the desired number of clusters is achieved. Besides the weaknesses inherited from K-means, such as sensitivity to initialization and noise, Bisecting K-means relies on the order of splitting and a variance-based splitting criterion.

The objective of Bisecting K-means is to minimize the within-cluster sum of squares (WCSS) for each split. Formally, if C is a cluster containing points $\{x_1, x_2, \ldots, x_n\}$, the within-cluster sum of squares (WCSS) is given by:

$$\text{WCSS}(C) = \sum_{i=1}^{|C|} \|x_i - \mu_C\|^2 \tag{5}$$

where μ_C is the centroid of cluster C, and $\|x_i - \mu_C\|$ is the Euclidean distance between point x_i and the centroid μ_C.

BiKmeans aims to minimize the total WCSS by hierarchically splitting the cluster C into two sub-clusters C_1 and C_2, such that:$\text{WCSS}(C) \approx \text{WCSS}(C_1) + \text{WCSS}(C_2)$. This hierarchical approach can result in better clustering for datasets with complex structures, especially in handling imbalanced clusters efficiently. The algorithm stops when the desired number of clusters is reached.

DBSCAN identifies clusters based on the density of data points using two parameters: ϵ (the maximum radius of a neighborhood) and minPts (the minimum number of points required to form a dense region) [7]. A point x is a core point if at least minPts points are within its ϵ-neighborhood:

$$N_\epsilon(x) = \{y \in D \mid \|x - y\| \leq \epsilon\} \tag{6}$$

where D is the input data. Clusters are formed by recursively connecting density-reachable points.

To cluster a given point, DBSCAN first examines its ϵ-neighborhood. If the neighborhood contains at least minPts points, the point is marked as a core point and a new cluster is initialized. Otherwise, the point is marked as an outlier. Points within the ϵ-neighborhood of a core point are added to the cluster and this process continues until all reachable points are assigned to clusters. Points that do not belong to any cluster are marked as outliers. DBSCAN can detect arbitrarily shaped clusters and outliers but may struggle with clusters of varying density.

HDBSCAN is an advanced version of DBSCAN. It works by identifying densely packed areas of data points and building a hierarchy of clusters from the most to the least dense [8]. It then selects the best clusters based on this hierarchy and ignores outliers. This process allows HDBSCAN to find natural groupings in the data, even if the clusters have different shapes and sizes, and to automatically identify and exclude outliers, making it a powerful and user-friendly clustering algorithm. Additionally, HDBSCAN can automatically determine the best parameters, making it easier to use.

OPTICS is another density-based clustering algorithm similar to DBSCAN but more robust to varying densities [9]. Imagine the data as a landscape with hills and valleys, where hills represent dense areas and valleys represent sparse areas. OPTICS "walks" through this landscape, visiting each data point and recording the distance to its nearest neighbor and the local density. This information is used to create a reachability plot, which maps out the density variations across the data points.

By analyzing the reachability plot, OPTICS can identify clusters as dense regions (hills) and separate them from sparse regions (valleys) [9]. Unlike some other algorithms, OPTICS can handle clusters with varying densities and doesn't require a fixed density threshold. This makes it particularly useful for complex datasets where clusters have different shapes and densities, allowing for a more flexible and detailed clustering analysis.

2.3   Clustering Evaluation Metrics

We employ three metrics to evaluate clustering performance under different normalization methods. The first two metrics, AMI and D-index, assume the ground truth is available in the final clustering. The third metric, the Silhouette score, evaluates clustering performance when the ground truth cannot be used, even if it is available. For example, ground truth information may not be used to evaluate the final clustering performance when data is imbalanced because the data imbalance may prevent building an effective mapping to align the clustering indices with the ground truth labels in the same coordinate system.

Adjusted Mutual Information (AMI) evaluates the similarity between two clusterings, adjusting for chance [10]. It measures the mutual information (MI) between a ground truth clustering U and a clustering algorithm's result V, quantifying the amount of shared information. The AMI adjusts this by comparing it to the expected MI (EMI) of random clusterings. The AMI is calculated as:

$$\mathrm{AMI}(U, V) = \frac{\mathrm{MI}(U, V) - \mathrm{EMI}(U, V)}{\max(H(U), H(V)) - \mathrm{EMI}(U, V)} \tag{7}$$

where $H(U)$ and $H(V)$ are the entropies of the clusterings. This adjustment ensures that AMI ranges from 0 (no agreement better than random) to 1 (perfect agreement), making it a robust and normalized measure for clustering comparison, especially for imbalanced clusters.

d-index is an interpretable assessment metric ranging from 0 to 2 for clustering and classification [11]. A high d-index (e.g., 1.8) indicates better clustering. The multiclass d-index is defined as

$$d = \frac{1}{K} \sum_{i=1}^{K} \left(\log_2(1 + \alpha_i) + \log_2\left(1 + \frac{s_i + p_i}{2}\right) \right) \tag{8}$$

where α_i is accuracy, s_i is specificity, and p_i is precision for class i among $K = \{0, 1, 2, \ldots, k - 1\}$ classes.

This metric can effectively evaluate the clustering model's ability to handle input data and overcome possible evaluation biases caused by imbalanced data.

The Silhouette score evaluates clustering quality by measuring how well a data point fits within its assigned cluster compared to other clusters [12]. For each data point i, the Silhouette score $s(i)$ is calculated using the average distance $a(i)$ to points within the same cluster and the minimum average distance $b(i)$ to points in any other cluster. The formula is:

$$s(i) = \frac{b(i) - a(i)}{\max(a(i), b(i))} \tag{9}$$

A score close to 1 indicates well-clustered points, close to 0 indicates boundary points, and close to -1 indicates misclassified points.

The overall Silhouette score for the dataset is the average of $s(i)$ for all points, providing a single metric to assess clustering quality. Higher average Silhouette scores signify better-defined clusters, aiding in determining the optimal number of clusters and improving clustering algorithms.

3   Results

This section presents benchmark datasets, which are classified as well-structured, extremely imbalanced, and lightly imbalanced data involved in this study, as well as the partitioning and density-based clustering results on them under different normalization methods.

3.1   Data

We include three benchmark datasets: Iris, Android malware, and Breast cancer. The Iris dataset is a well-structured balanced dataset, while the Android malware and Breast cancer datasets are both imbalanced. The Android malware dataset is an extremely imbalanced one with a majority ratio of 96.67% and the Breast Cancer dataset is a slightly imbalanced one with a majority ratio of 62.74%.

The Iris dataset is a classic dataset in machine learning, consisting of 150 samples of iris flowers from three species: Setosa, Versicolor, and Virginica, with 50 samples per species. It includes four numerical features: sepal length, sepal width, petal length, and petal width, all measured in centimeters.

The Andriod malware dataset has 510 observations, consisting of 17 malware samples and 493 normal apps, across 41 features. The features include BatteryVoltage, cpuUsage, battery temperature, memoryFilePages, memoryFreePages, memoryWritebackPages, binderTransaction etc. This diverse set of features captures different aspects of app behavior, making the dataset valuable for analyzing and distinguishing between malicious and normal applications.

The Breast Cancer dataset, also known as the Wisconsin Diagnostic Breast Cancer (WDBC) dataset, consists of 569 observations with 30 numerical features [13]. These features, including measurements such as radius, texture, perimeter, area, smoothness, compactness, concavity, concave points, symmetry, and fractal dimension, capture various properties of cell nuclei. Each observation is labeled as 212 malignant samples and benign 357 samples. Figure 1 illustrates the three datasets via principal component analysis (PCA) visualization. It is easy to see that versicolor and virginica samples are more likely to produce mis-clustering cases due to their partial mixture. Both of them are imbalanced datasets, with the Android dataset being extremely imbalanced. This extreme imbalance usually prevents the use of ground truth information in clustering evaluation.

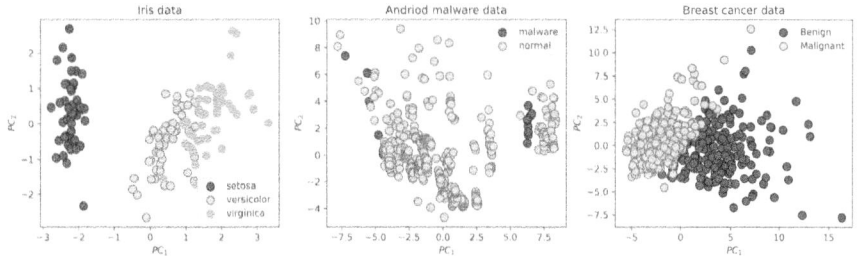

Fig. 1. Data visualization of three datasets via principal component analysis

3.2   Well-Structured Data Clustering

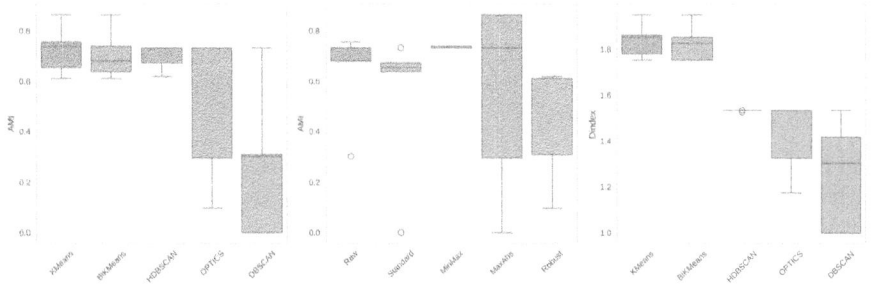

Fig. 2. Comparisons of partitioning and density-based clustering methods under different normalization methods on Iris data

Figure 2 compares the performance of partitioning and density-based clustering methods across different normalization methods: standard, minmax, maxabs, and robust scaling, in terms of AMI and d-index values on Iris data clustering. As mentioned before, the dataset is well-structured with balanced label distributions and possesses a 'ready for clustering' property even in its raw form. Deciphering the impacts of normalization methods on partitioning and density-based clustering can answer the important question: 'How are clustering methods impacted by different normalization techniques for a well-structured dataset?

It shows that partitioning methods are less impacted by normalization than density-based methods on this well-structured dataset while maintaining superior performance. This may imply that partitioning methods are more applicable to well-structured data than density-based methods due to their robustness to normalization and better performance.

Partitioning Methods Outperform Density-Based Methods. The left sub-figure shows that K-means and BiKmeans have higher median AMI values with relatively smaller IQRs, indicating consistently leading clustering performance compared to the density-based methods. DBSCAN and OPTICS show greater AMI variability, with DBSCAN having the lowest median and widest IQR. HDBSCAN only outperforms DBSCAN and OPTICS.

DBSCAN is Most Impacted by Normalization. The right sub-figure echoes these results with d-index values. K-means and BiKmeans have the highest d-index values with small IQRs, although BiKmeans has a slightly lower median. HDBSCAN and OPTICS have moderate d-index values, with OPTICS showing a higher median. DBSCAN has the lowest median d-index and largest IQR, indicating inconsistent clustering quality and being the most impacted by different normalization methods. It is noted that the parameters of DBSCAN are fine-tuned as the other parameters in the other density-based methods.

Minmax Normalization Matches Well with Most Methods. The middle sub-figure shows the performance of clustering methods using raw data, standard scaling, minmax scaling, maxabs scaling, and robust scaling. Minmax scaling outperforms the others, including standard scaling, with a higher median AMI value. Maxabs scaling has the highest median AMI but greater variability, indicating it may suit only specific methods like K-means. Raw data exhibits the lowest and most variable AMI values. Robust scaling shows moderate performance with some variability.

The reasons for the superiority of partitioning methods to density-based methods lies in the following aspects. 1) Well-structured data often exhibits clear, distinct clusters. Partitioning methods, such as k-means are generally well-suited for datasets with clear structure and balanced label distributions. 2) Partitioning methods assume that clusters are roughly equal in size and shape, which aligns well with the characteristics of well-structured data. In contrast, density-based methods, such as DBSCAN, are designed to find clusters of varying shapes and densities, which might not be necessary for well-structured data like Iris data. 3) Density-based methods excel in identifying noise and outliers in datasets with varying densities. However, in well-structured datasets, where noise and outliers are minimal, the added complexity of density-based methods may not provide good benefits over partitioning methods.

3.3 Extremely Imbalanced Data Clustering

We find that density-based clustering methods are more heavily impacted by different normalization methods when dealing with imbalanced data, especially extremely imbalanced data. As we mentioned before, we employ silhouette scores to evaluate clustering performance because ground truth information cannot be used to evaluate the final clustering performance for such an extremely imbalanced dataset.

Density-based methods demonstrate instability or even unreliability under the Android malware dataset under different normalization methods. For example, DBSCAN mis-clusters all malware samples as normal ones under the minmax and maxabs normalizations. HDBSCAN reports five classes under the minmax normalization with a silhouette score of 0.18, but it cannot predict any instances in classes 0, 3, or 4, suggesting unreliable performance, especially since it reports a silhouette score of 0.64 on the raw data. Similarly, OPTICS reports a silhouette score of -0.26 under robust normalization and 0.27 under minmax normalization.

In contrast, partitioning methods demonstrate obvious advantages over density-based methods in terms of silhouette scores. It is also interesting to note that fine-tuned parameters may not contribute to enhancing imbalanced data clustering for density-based methods. Figure 3 compares the silhouette scores of partitioning and density-based clustering methods under different normalization methods. The left sub-figure shows that K-means and BiKmeans outperform the density-based methods under maxabs, minmax, and robust normalization. Robust scaling seems to produce excellent clustering results for K-means and BiKmeans, achieving a silhouette score of 0.98, while HDBSCAN and OPTICS report silhouette scores of 0.0 and -0.26, respectively.

The right sub-figure illustrates that partitioning methods, K-means and BiKmeans, are much less impacted by normalization compared to their density-based counterparts, achieving leading clustering performance. DBSCAN has a lower median silhouette score compared to K-means and BiKmeans but outperforms HDBSCAN and OPTICS. OPTICS has the lowest median silhouette score with a narrow interquartile range (IQR) and a few outliers. In other words, the density-based methods (e.g., DBSCAN) either show mediocre performance or poor and unreliable performance, such as OPTICS and HDBSCAN.

Why Partitioning Methods Are More Robust on Extremely Imbalanced Data? Partitioning methods, such as K-means and BiKmeans, are more robust than density-based methods in handling extremely imbalanced data for several reasons: 1) *Assumption of spherical clusters*: Partitioning methods assume clusters are spherical and roughly equal in size [14]. This helps maintain stability even with imbalanced data, as the method partitions data into predefined shapes and sizes. 2) *Density thresholds*: Density-based methods, like DBSCAN, HDBSCAN, and OPTICS, do not assume specific cluster shapes or sizes. They rely on density thresholds, which can lead to instability with imbalanced data since low-density areas representing minority classes might be mis-

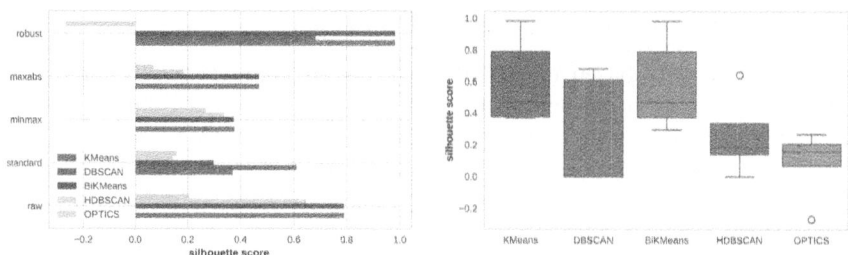

Fig. 3. Silhouette score comparisons of partitioning and density-based clustering methods under different normalization methods on the extremely imbalanced Andriod dataset

classified or ignored. 3) *Normalization sensitivity*: Density-based methods are highly sensitive to normalization because density calculations vary significantly with different scales. This sensitivity is exacerbated in imbalanced datasets. 4) *Parameter sensitivity*: Density-based methods require fine-tuning parameters, which can be biased in imbalanced situations, leading to merging distinct clusters or fragmenting single clusters into multiple parts.

3.4 Lightly Imbalanced Data Clustering

The Breast Cancer dataset is lightly imbalanced due to its relatively low majority ratio of 62.74%. The ground truth information can be employed in clustering performance evaluation. However, we also include AMI, d-index, and silhouette scores to evaluate the clustering performance for this dataset to avoid possible biases from data imbalance (Fig. 4). Our results show that partitioning methods still have advantages over their density-based counterparts, although both are significantly impacted by standard and robust scaling. For example, all methods perform poorly under standard and robust scaling. K-means, DBSCAN, and OPTICS achieve zero AMI, while BiKmeans and HDBSCAN only achieve AMI values of 0.13 and 0.06, respectively. Similarly, DBSCAN and OPTICS achieve

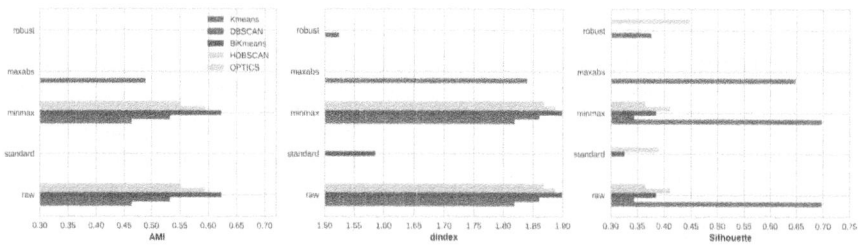

Fig. 4. AMI, d-index, and silhouette score comparisons of partitioning and density-based clustering methods under different normalization methods on the Breast Cancer dataset

AMI values of 0.08 and 0.04, respectively, while the other three methods have zero AMI values. This is because the Breast Cancer dataset does not follow a normal distribution and is homogeneous without outliers. Using these two normalization methods may seriously distort the data and result in negative or very poor clustering performance.

BiKmeans achieves the leading AMI, d-index, and silhouette values under raw data and minmax normalized data among all methods. K-means is the only method with an AMI value of 0.49, a d-index value of 1.83, and a silhouette value of 0.65 under maxabs scaling, while the others have zero AMI, d-index, and silhouette values. K-means also has the best silhouette value under raw data and minmax normalized data.

The reasons why partitioning methods outperform density-based methods under different normalization techniques are similar to those for extremely imbalanced data: the former assumes spherical clusters that align with most data in clustering, while density-based methods can be too sensitive to parameters. Fine-tuning these parameters may not yield desirable results when the data is imbalanced. In many scenarios, they can even amplify the negative impacts of data imbalance in clustering.

4   Discussion

The current study has the following limitations. 1) Our datasets are limited, and more benchmark datasets should be included in this study to extend the scope of our research. 2) We only include four commonly used normalization methods. Other normalization methods, such as the Power Transformer or more complex ones, are not included in this work [15]. 3) More partitioning and density-based clustering models such as GMM and BIRCH (Balanced Iterative Reducing and Clustering using Hierarchies) should be included in this study to gain a comprehensive understanding of how normalization impacts different clustering models [16]. 4) We employ elbow point search followed by heuristic search to fine-tune the parameters in the density-based methods (e.g., minpts) for both well-structured data like the Iris dataset and imbalanced data [8,17]. More tailored fine-tuning methods should be employed for imbalanced data in density models, though there is currently no work on this topic. We plan to address these limitations in future work, and we are particularly interested in investigating more refined fine-tuning techniques for density-based methods on imbalanced data [18].

An interesting question that remains to be answered is whether we should seek tailored normalization methods not only for input data but also for different clustering methods. Some normalization methods may scale data well, but they may not be compatible with certain clustering methods, particularly density-based methods. This is mainly because clustering methods have varying assumptions and sensitivities that are not fully considered in the current design of normalization methods. For instance, density-based methods like DBSCAN are more sensitive to density variations and may benefit from normalization techniques that emphasize these variations. Furthermore, imbalanced data can significantly impact clustering results. Tailored normalization methods can help address issues caused by imbalance, ensuring that the clustering algorithm does not become biased towards the dominant class.

We plan to address this problem in our future work by seeking tailored min-max-like scaling for input data and selected clustering methods whiling miltigating its possible biases. According to our experiments, min-max scaling appears to be a good choice for almost all data and clustering methods due to its data-driven nature, although it may introduce some bias for certain data.

5   Conclusion

Our study finds that partitioning methods demonstrate greater efficiency and robustness under different normalization methods compared to density-based methods. This suggests that partitioning methods can be a good candidate for clustering well-structured and imbalanced data. Contrary to the general assumption that density-based methods are robust to normalization due to their local-level density-based clustering analysis, our study finds that the spherical cluster assumption of partitioning methods aligns well with well-structured or even imbalanced data. The sensitivity to parameters and the need for parameter fine-tuning can be a challenge for handling imbalanced data, which may prevent density-based methods from performing well with such data. Our study suggests that density-based methods can be highly sensitive to normalization, whether dealing with well-structured or extremely imbalanced data. Special normalization techniques customized for different density-based methods should receive more attention in investigations, alongside efforts to enhance the efficiency and stability of density-based methods in imbalanced data clustering.

Furthermore, we are interested in investigating the impacts of different normalization techniques on partitioning and density-based methods in imbalanced data clustering when resampling or data augmentation techniques are employed to handle data imbalance [19]. It is expected that partitioning methods may still lead over their density-based peers, though more information and evidence need to be collected. Such research can also shed more light on how to handle the imbalanced data clustering problem, which remains unsolved in various data science domains [20].

Acknowledgment. We would like to express our deep gratitude to Dr. Henry Han for his mentoring and help with this project. We also thank the anonymous reviewers for their insightful suggestions and comments that contributed to enhancing this paper.

References

1. Singh, D., Singh, B.: Investigating the impact of data normalization on classification performance. Appl. Soft Comput. **97**, 105524 (2020)
2. Han, H., Men, K.: How does normalization impact RNA-SEQ disease diagnosis? J. Biomed. Inform. **85**, 80–92 (2018)

3. Zass, R., Shashua, A.: Doubly stochastic normalization for spectral clustering. In: Advances in Neural Information Processing Systems, vol. 19 (2006)
4. Viswanathan, V., Gashteovski, K., Lawrence, C., Wu, T., Neubig, G.: Large language models enable few-shot clustering. Trans. Assoc. Comput. Linguist. **12**, 321–333 (2024)
5. Jian-Wei, L., Hui-Dan, Z., Xiong-Lin, L., Jun, X.: Research progress on batch normalization of deep learning and its related algorithms. Acta Automatica Sinica **46**(6), 1090–1120 (2020)
6. Rokach, L., Maimon, O.: Clustering methods. In: Data Mining and Knowledge Discovery Handbook, pp. 321–352 (2005)
7. Kriegel, H.P., Kröger, P., Sander, J., Zimek, A.: Density-based clustering. Wiley Interdisc. Rev. Data Mining Knowl. Discov. **1**(3), 231–240 (2011)
8. Campello, R.J., Moulavi, D., Zimek, A., Sander, J.: Hierarchical density estimates for data clustering, visualization, and outlier detection. ACM Trans. Knowl. Discov. Data (TKDD) **10**(1), 1–51 (2015)
9. Ankerst, M., Breunig, M.M., Kriegel, H.P., Sander, J.: OPTICS: ordering points to identify the clustering structure. ACM SIGMOD Rec. **28**(2), 49–60 (1999)
10. Vinh, N.X., Epps, J.: Information theoretic measures for clusterings comparison: variants, properties, normalization and correction for chance. J. Mach. Learn. Res. **11**, 2837–2854 (2010)
11. Han, H., Wu, Y., Wang, J., Han, A.: Interpretable machine learning assessment. Neurocomputing **561**, 126891 (2023)
12. Rousseeuw, P.J.: Silhouettes: a graphical aid to the interpretation and validation of cluster analysis. J. Comput. Appl. Math. **20**, 53–65 (1987)
13. https://archive.ics.uci.edu/dataset/17/breast+cancer+wisconsin+diagnostic
14. Hadi, A.S.: A new distance between multivariate clusters of varying locations, elliptical shapes, and directions. Pattern Recogn. **129**, 108780 (2022)
15. Lima, F.T., Souza, V.M.: A large comparison of normalization methods on time series. Big Data Res. **34**, 100407 (2023)
16. Zhang, T., Ramakrishnan, R., Livny, M.: BIRCH: an efficient data clustering method for very large databases. ACM SIGMOD Rec. **25**(2), 103–114 (1996)
17. Schubert, E.: Stop using the elbow criterion for k-means and how to choose the number of clusters instead. ACM SIGKDD Explor. Newsl. **25**(1), 36–42 (2023)
18. Xu, Z., Shen, D., Nie, T., Kou, Y., Yin, N., Han, X.: A cluster-based oversampling algorithm combining SMOTE and k-means for imbalanced medical data. Inf. Sci. **572**, 574–589 (2021)
19. Salehi, A.R., Khedmati, M.: A cluster-based SMOTE both-sampling (CSBBoost) ensemble algorithm for classifying imbalanced data. Sci. Rep. **14**(1), 5152 (2024)
20. Wen, G., Li, X., Zhu, Y., Chen, L., Luo, Q., Tan, M.: One-step spectral rotation clustering for imbalanced high-dimensional data. Inf. Process. Manag. **58**(1), 102388 (2021)

Quantitative Stock Market Modeling Using Multivariate Geometric Random Walk

Michael Pokojovy[1]([✉])[iD], Andrews T. Anum[2][iD], Obed Amo[1][iD],
Maria C. Mariani[3][iD], and Michael C. Orosz[4][iD]

[1] Old Dominion University, Norfolk, VA 23529, USA
{mpokojovy,oamo}@odu.edu
[2] University of New England, Biddeford, ME 04005, USA
aanum@une.edu
[3] The University of Texas at El Paso, El Paso, TX 79968, USA
mcmariani@utep.edu
[4] GECU Federal Credit Union, El Paso, TX 79925, USA
mike.orosz@gecu.com

Abstract. We propose a new stock market model based on multivariate geometric random walk without imposing any parametric assumptions (such as Gaussianity) or structural assumptions (such as ellipticity) on log-increments and establish connection with continuous-time Geometric Brownian Motion (GBM). Our approach can be applied to simultaneous modeling of a variety of stocks traded at multiple stock exchanges and can adequately account for heavy tails and other distributional departures from Gaussianity. Both calibration and forecasting steps assume a nonparametric innovation process. Model calibration involves multivariate imputation to account for partially overlapping trading hours, while simulation and forecasting is performed using resampling from imputed observations. Our model is applied to a wide selection of stocks traded at NYSE and LSE. Analyzing three months' worth of closing stock price data collected from 324 stocks sampled every minute, respective forecast regions are constructed and successfully backtested on a one-month horizon of historical data. Based on a benchmarking study, our model was shown to outperform standard GBM as well as geometric random walk models with Student's t-like and Laplace log-increments at forecasting future value of a custom portfolio comprised of US and UK stocks.

Keywords: Missing values · Imputation · Forecasting · Risk management · t-SNE

1 Introduction

In the dynamic world of financial markets, the ability to model and predict stock price movements is of paramount importance for researchers, investors,

This work received partial funding support from National Science Foundation [DMS-2210929, ERC-ASPIRE-1941524, DUE-2216396] and Department of Education [Award #P116S210004].

H. Han (Ed.): SDSC 2024, CCIS 2158, pp. 48–63, 2024.
https://doi.org/10.1007/978-3-031-67871-4_4

portfolio managers and policymakers alike. Traditional models of stock price dynamics, such as the Black-Scholes model and the Capital Asset Pricing Model (CAPM) [7,13,26], have laid the foundation for modern financial theory. These models, which often assume log-normal distributions of returns and linear relationships between risk and return, have been instrumental in the development of derivative pricing, portfolio optimization and risk management strategies. However, empirical studies have consistently shown that stock returns exhibit features such as heavy tails, volatility clustering and non-linear dependencies that are not adequately captured by these classical models [9,21].

To address these limitations, recent research has increasingly focused on developing more sophisticated models that can capture complex statistical properties of financial time series. Among these, the Geometric Brownian Motion (GBM) model remains a cornerstone in continuous-time finance, providing a simple yet powerful framework for modeling stock prices [23]. Despite wide popularity, the GBM model's assumption of constant volatility and normally distributed returns has been criticized for its inability to account for the empirical irregularities observed in financial markets [4]. These challenges are further compounded by the difficulties faced by Gaussian Maximum Likelihood Estimation (MLE) in high-dimensional portfolio optimization settings [22]. It was shown that estimation errors, particularly in determining the expected rates of return and the variance-covariance matrix, can significantly undermine the performance of optimized portfolios. These errors can lead to degenerate investment strategies that either overly favor risk-free assets or result in severely diminished expected utility values, underscoring the need for models that can better accommodate the realities of high-dimensional financial data.

In response to these challenges, our work proposes a new stock market model based on a multivariate geometric random walk, eschewing restrictive parametric assumptions (such as Gaussianity) and structural assumptions (like ellipticity) inherent in prior models. This approach not only offers a more flexible representation of return distributions, accounting for heavy tails and other deviations from normality, but also establishes a crucial connection with continuous-time GBM, bridging the gap between discrete and continuous modeling efforts.

The significance of this work extends into the domain of risk management. Traditional risk management tools, such as value at risk (VaR) and conditional value at risk (CVaR) measures as well as stress testing, rely heavily on the underlying assumptions about the distribution of returns [1,15,16,24]. CVaR, also known as expected shortfall, is a risk measure that quantifies the expected loss in scenarios that are worse than the level of loss specified by the VaR, providing more accurate assessment of potential extreme losses [24]. This measure is crucial for investors seeking to understand the full spectrum of downside risk, enabling more informed risk management and investment decisions. Stress testing in the context of structural credit portfolio models involves evaluating credit portfolios under adverse economic conditions by translating stress scenarios into constraints on risk factors [16]. This is achieved by conditioning on the range of values that risk factors may attain through truncation, thereby formalizing the constraints and allowing for an empirical examination of the stress impact on asset correlations and default probabilities. This process helps identify potential vulnerabilities within investment strategies, offering insights for developing more robust risk mitigation approaches.

By adopting a semiparametric approach, our model provides a more reliable foundation for estimating risk measures, accounting for the extreme events and anomalies that are often observed in financial markets. This advancement proves essential for risk management, where modeling and forecasting of stock prices play an important role in estimating and mitigating risks across various investment strategies. Moreover, the construction of per-comparison and simultaneous prediction regions for multiple stocks, along with prediction intervals for stock portfolios, offers a powerful tool for assessing the risks associated with diverse investment strategies. Our model's development reflects a recent paradigm shift in financial modeling towards a more comprehensive and accurate framework. This transition is underpinned by recent developments in XVA (Valuation Adjustment) frameworks, which address the multifaceted nature of risk, including market, credit, funding, and capital considerations [12,25]. This underscores the need for models that can not only capture the complexities of financial markets but also provide actionable insights for risk management.

Machine learning techniques, including various non-linear autoregressive models based on artificial neural networks (ANN), random forests (RF) and support vector machines (SVM), have become increasingly popular for predicting stock market dynamics due to their flexibility in capturing complex patterns in time series [2,3,20]. These methods leverage historical data to forecast future stock prices, each offering distinct advantages and facing specific challenges. ANN, for example, require extensive data for training to avoid overfitting, while SVM often struggle with high dimensionality prevalent in financial datasets. Traditional recurrent ANN can also suffer from the vanishing gradient problem, which makes it challenging for them to capture long-term dependencies. Long Short-Term Memory (LSTM) [14] networks and Gated Recurrent Unit (GRU) networks [8] are variants of RNN that are better equipped to capture long-term dependencies by incorporating gating mechanisms. LSTM and GRU networks have been widely used for nonlinear autoregressive time series modeling due to their ability to capture both short-term and long-term dependencies in the data.

Another major challenge faced by ML methodologies is missing data, a frequent issue in stock markets due to varying global trading hours. While univariate RF allow for missing values in predictors [29], no comparable imputation instruments are currently available for multivariate RF models [27]. As for ANN, a deep learning model referred to as Gated Recurrent Unit with Missing Data (GRU-D) was proposed in [6] which incorporates two representations of missing patterns, viz. masking and time intervals, into the architecture of a GRU-based recurrent ANN. The work focused on classification of time series with missing values. The empirical experiments presented show that the method is suitable for many time series classification problems with missing data. A Forward and Backward Variable-Sensitive LSTM (FBVS-LSTM) model was introduced by [11] to specifically address the problem of missing-not-at-random (MNAR) data in environmental data science where sensor reliability and data completeness are major issues. Forecasting of meteorological multivariate time series, like air quality, was considered in [11]. The work focused on multivariate time series data and has not yet been tested in high-dimensional settings. To circumvent these and other challenges with existing ANN, RF and SVM methodologies, we chose to impute the log-differenced stock prices with nonparametric RF instead of directly training an autoregressive ML model.

The rest of the paper is structured as follows. Section 2 introduces the concept of GMB, setting the stage for the introduction of our semi-parametric extension based on a non-parametric random walk methodology. In Sect. 3, we elaborate on the estimation process for stock exchanges that operate continuously, as well as those with partially overlapping trading hours. Section 4 discusses how our technique can be used to simulate and forecast future trajectories of stock prices. In Sect. 5, we demonstrate the practical application of our methodology to a comprehensive selection of stocks traded on the New York Stock Exchange (NYSE) and the London Stock Exchange (LSE). Analyzing three months of minute-by-minute closing stock price data, we generate respective forecast regions and backtest them on a one-month horizon. Based on a benchmarking study in application to forecasting the future value of a custom portfolio, we further show that our method outperforms both the Gaussian GBM model and parametric geometric random walks with multivariate Laplace and Student's t-like log-increments. The results show that the model can predict accurately, proving it is a useful tool for understanding the stock market. The paper concludes with Sect. 6, offering a summary and discussion of the findings and the implications of our proposed method.

2 Semiparametric Geometric Brownian Motion

Consider the stochastic differential equation (SDE) for a p-variate GBM $(\boldsymbol{S}_t)_{t\geq 0}$ with constant drift $\boldsymbol{a} \in \mathbb{R}^p$ and volatility matrix $\boldsymbol{B} \in \mathbb{R}^{p\times p}$:

$$\mathrm{d}\boldsymbol{S}_t = \mathrm{diag}\left(\boldsymbol{S}_t\right)\left(\boldsymbol{a}\mathrm{d}t + \boldsymbol{B}\mathrm{d}\boldsymbol{W}_t\right) \text{ for } t > 0, \quad \boldsymbol{S}_0 = \boldsymbol{\xi}. \qquad (2.1)$$

Here, $(\boldsymbol{W}_t)_{t\geq 0}$ is a p-variate standard Brownian motion and $\boldsymbol{\xi} \in \mathbb{R}^p$ is a vector of known initial prices. Itô's calculus yields the solution process [19, Section 2]

$$\boldsymbol{S}_t = \mathrm{diag}\left\{\exp\left(t(\boldsymbol{a} - \tfrac{1}{2}\mathrm{diag}(\boldsymbol{B}\boldsymbol{B}')) + \boldsymbol{B}\boldsymbol{W}_t\right)\right\}\boldsymbol{\xi} \text{ for } t \geq 0. \qquad (2.2)$$

Introducing an equispaced time lattice $T = \{t_0, t_1, \ldots, t_k, \ldots\}$ with a step size $\Delta t \equiv t_k - t_{k-1} > 0$, Eq. (2.2) can be approximated via

$$\boldsymbol{S}_{t_k} = \mathrm{diag}\left\{\exp\left(t_k(\boldsymbol{a} - \tfrac{1}{2}\mathrm{diag}(\boldsymbol{B}\boldsymbol{B}')) + \boldsymbol{B}\sum_{l=1}^{k}(\Delta t)^{1/2}\boldsymbol{\epsilon}_l\right)\right\}\boldsymbol{\xi} \text{ for } k \geq 0 \quad (2.3)$$

where the Wiener process $(\boldsymbol{W}_t)_{t \geq 0}$ is approximated via Gaussian "random walk"

$$\boldsymbol{W}_{t_k} \approx \boldsymbol{W}_{t_k}^{\Delta t} \equiv (\Delta t)^{1/2} \sum_{l=1}^{k} \boldsymbol{\varepsilon}_l \quad \text{with } \boldsymbol{\varepsilon}_l \overset{\text{i.i.d.}}{\sim} \mathcal{N}(\boldsymbol{0}_p, \boldsymbol{I}_{p \times p}).$$

By induction, Eq. (2.3) can be expressed in the recursive form:

$$\boldsymbol{S}_{t_k} = \text{diag}\left\{ \exp\left(\Delta t_k \left(\boldsymbol{a} - \tfrac{1}{2}\text{diag}(\boldsymbol{B}\boldsymbol{B}')\right) + (\Delta t)^{1/2}\boldsymbol{B}\boldsymbol{\varepsilon}_k \right) \right\} \boldsymbol{S}_{t_{k-1}}, \ k \geq 1. \tag{2.4}$$

We refer to Eqs. (2.3) and (2.4) as Gaussian geometric random walk model. This model is widely used in quantitative finance (cf. [23] and references therein). Being parametric in nature, Eqs. (2.3) and (2.4) are well amenable to classical statistical analysis and efficient use in Monte-Carlo simulations. A common empirical observation made by practitioners is though that the log-returns are oftentimes not Gaussian, but rather exhibit heavier tails. Moreover, the underlying distribution may even fail to be elliptic if the log-price differences for various stocks happen to have tails with diverse asymptotic behavior.

We propose a new flexible way to extend Eq. (2.4) by replacing the Gaussian random walk in Eq. (2) with its nonparametric counterpart

$$\boldsymbol{W}_{t_k}^{\Delta t} := (\Delta t)^{1/2} \sum_{l=1}^{k} \boldsymbol{\varepsilon}_l \text{ with } \boldsymbol{\varepsilon}_l \overset{\text{i.i.d.}}{\sim} F_0(\cdot), \tag{2.5}$$

where $F_0(\cdot)$ is the cumulative distribution function (cdf) of a square-integrable p-variate random vector with mean vector $\boldsymbol{0}_p$ and covariance matrix $\boldsymbol{I}_{p \times p}$. No parametric assumptions are made on $F_0(\cdot)$. Thus, we arrive at

$$\boldsymbol{S}_{t_k} = \text{diag}\left\{ \exp\left(\Delta t_k \left(\boldsymbol{a} - \tfrac{1}{2}\text{diag}(\boldsymbol{B}\boldsymbol{B}')\right) + (\Delta t)^{1/2}\boldsymbol{B}\boldsymbol{\varepsilon}_k \right) \right\} \boldsymbol{S}_{t_{k-1}}, \ k \geq 1 \tag{2.6}$$

with $\boldsymbol{S}_{t_0} = \boldsymbol{\xi}$ and $\boldsymbol{\varepsilon}_k \overset{\text{i.i.d.}}{\sim} F_0(\cdot)$ as a semiparametric version of Eq. (2.4).

While $F_0(\cdot)$ trivially includes such classic cases as multivariate Gaussian or, more generally, Student's t_ν-distribution, it also naturally allows for any elliptic or non-elliptic location-scale family. For example, given a cdf $F(\cdot)$ underlying some square-integrable distribution, the standardized baseline c.d.f.

$$dF_0(\boldsymbol{z}) = |\boldsymbol{\Sigma}|^{1/2} dF\left(\boldsymbol{\mu} + \boldsymbol{\Sigma}^{1/2}\boldsymbol{z}\right) \text{ for } \boldsymbol{z} \in \mathbb{R}^p \tag{2.7}$$

with $\boldsymbol{\mu} = \int_{\mathbb{R}^p} \boldsymbol{x} dF(\boldsymbol{x})$, $\boldsymbol{\Sigma} = \int_{\mathbb{R}^p} (\boldsymbol{x} - \boldsymbol{\mu})(\boldsymbol{x} - \boldsymbol{\mu})' dF(\boldsymbol{x})$ can be computed, which gives rise to a location-scale family

$$\left(dF_{\boldsymbol{\mu}, \boldsymbol{\Sigma}}(\cdot)\right)_{\boldsymbol{\mu} \in \mathbb{R}^p, \boldsymbol{\Sigma} \in \text{SPD}(p \times p)} \text{ with } dF_{\boldsymbol{\mu}, \boldsymbol{\Sigma}}(\cdot) \equiv |\boldsymbol{\Sigma}|^{-1/2} dF_0\left(\boldsymbol{\Sigma}^{-1/2}(\cdot - \boldsymbol{\mu})\right). \tag{2.8}$$

This concept extends beyond square-integrable distributions when the mean vector and the covariance matrix are replaced with robust counterparts such as Rousseeuw's MCD location and scatter [5]. It may superficially appear that the nonparametric random walk in Eq. (2.5) has the downside that the family $(dF_{\mu,\Sigma})_{\mu,\Sigma}$ may fail to be a Lévy α-stable distribution [18] and, thus, not linearly closed. Nonetheless, this theoretical "paradox" can only occur in our model if the log-returns violate the condition at the population level. While it remains debatable if the empirically observed distribution should always be forced into the "bed of Procrustes" of α-stable distributions even at the price of reduced predictive power, our model neither enforces nor destroys this property.

3   Model Estimation

Consider a stock market comprised of m stock exchanges $SE_1, \ldots, SE_m$ trading $p_1, p_2, \ldots, p_m$ stocks, respectively. Denoting $\sum_{i=1}^{m} p_i = p$, let $S_t^1, S_t^2, \ldots, S_t^p$ be the prices of the p stocks traded at $SE_1, \ldots, SE_m$. Further, let $I_i \subset \{1, 2, \ldots, p\}$ be the indices of the p_i stocks traded at SE_i for $i = 1, 2, \ldots, m$. Assume the p-variate price vector $\boldsymbol{S}_t = (S_t^1, S_t^2, \ldots, S_t^p)$ follows the model (2.6) with some sampling frequency $\Delta t > 0$. Consider the discrete time grid $T_n = \{t_0, t_1, \ldots, t_n\}$, where $t_k = t_0 + (\Delta t)k$ for $k = 0, 1, \ldots, n$ are the sampling times.

3.1   Non-stop Operating Stock Exchanges

We begin with the basic situation when all stock prices are observable at any $t \in T_n$. Log-differencing Eq. (2.6), we obtain

$$\log(\boldsymbol{S}_{t_k}) - \log(\boldsymbol{S}_{t_{k-1}}) = \Delta t_k \left(\boldsymbol{a} - \tfrac{1}{2} \operatorname{diag}(\boldsymbol{B}\boldsymbol{B}')\right) + (\Delta t)^{1/2} \boldsymbol{B}\boldsymbol{\varepsilon}_k \text{ for } k = 1, 2, \ldots, n$$

where $\boldsymbol{\varepsilon}_k \overset{\text{i.i.d.}}{\sim} F_0(\cdot)$ and $F_0(\cdot)$ satisfies Eq. (2.7). Thus,

$$\log(\boldsymbol{S}_{t_k}) - \log(\boldsymbol{S}_{t_{k-1}}) \overset{\text{i.i.d.}}{\sim} F(\cdot) \text{ for } k = 1, 2, \ldots \text{ with}$$

with $dF(\cdot) = |\boldsymbol{\Sigma}|^{-1/2} F_0\left(\boldsymbol{\Sigma}^{-1/2}(\cdot - \boldsymbol{\mu})\right)$ where

$$\boldsymbol{\mu} := \Delta t \left(\boldsymbol{a} - \tfrac{1}{2} \operatorname{diag}(\boldsymbol{B}\boldsymbol{B}')\right), \quad \boldsymbol{\Sigma}^{1/2} := (\Delta t)^{1/2} \boldsymbol{B} \tag{3.1}$$

are the mean vector and root-covariance matrix. Letting $\boldsymbol{s}_t \in \mathbb{R}^p$ denote the observed stock price vector $\boldsymbol{S}_t$ value at time $t \in T_n$, the log-differences $\boldsymbol{x}_k := \log(\boldsymbol{s}_k) - \log(\boldsymbol{s}_{k-1})$ for $k = 1, 2, \ldots, n$ form an independent sample from $F(\cdot)$. Hence, a nonparametric estimate of $F(\cdot)$ can trivially be obtained from the dataset $\boldsymbol{x}_1, \ldots, \boldsymbol{x}_n$ via the standard empirical c.d.f.

$$\hat{F}_n(\cdot) = \frac{1}{n} \sum_{k=1}^{n} \mathbb{1}_{(-\infty, \boldsymbol{x}_k]}(\cdot) \quad \text{with} \quad \mathbb{1}_{(-\infty, \boldsymbol{x}]}(\boldsymbol{\xi}) = \begin{cases} 1, \ \xi_i \leq x_i \text{ for } 1 \leq i \leq p, \\ 0, \text{ otherwise.} \end{cases}$$

Further, the nonparametric plug-in estimates of $\boldsymbol{\mu}$ and $\boldsymbol{\Sigma}$ are given as

$$\hat{\boldsymbol{\mu}} = \int_{\mathbb{R}^p} \boldsymbol{x} d\hat{F}(\boldsymbol{x}) \equiv \frac{1}{n} \sum_{k=1}^{n} \boldsymbol{x}_k, \quad \hat{\boldsymbol{\Sigma}} = \int_{\mathbb{R}^p} (\boldsymbol{x} - \hat{\boldsymbol{\mu}})(\boldsymbol{x} - \hat{\boldsymbol{\mu}})' d\hat{F}_n(\boldsymbol{x}), \tag{3.2}$$

where the Bessel correction factor is usually employed in front of $\hat{\boldsymbol{\Sigma}}$ to render it unbiased. Therefore, in line with Eq. (3.1), $\boldsymbol{a}$ and $\boldsymbol{\Sigma}$ can be estimated via

$$\hat{\boldsymbol{a}} = (\Delta t)\hat{\boldsymbol{\mu}} + \tfrac{1}{2}\operatorname{diag}(\hat{\boldsymbol{B}}), \quad \hat{\boldsymbol{B}} = (\Delta t)^{1/2}\hat{\boldsymbol{\Sigma}}^{1/2} \tag{3.3}$$

and the nonparametric estimate of $\mathrm{d}F_0(\cdot)$ is $\mathrm{d}\hat{F}_0(\cdot) = |\hat{\boldsymbol{\Sigma}}|^{1/2}\mathrm{d}\hat{F}_n\big(\hat{\boldsymbol{\mu}} + \hat{\boldsymbol{\Sigma}}^{1/2}(\cdot)\big).$

3.2   Stock Exchanges with Partially Overlapping Trading Hours

It is more realistic though to assume that the stock exchanges are not operating non-stop and, possibly, have non-identical trading hours. While still imposing the model in Eq. (2.6), we assume that the data are observed only partially. For $i = 1, 2, \ldots, m$, let $T_n^i \subset T_n$ denote all time periods that fall within the trading hours of the i-th stock exchange SE_i. We assume that $T_n^i \neq \emptyset$ for $i = 1, 2, \ldots, m$ and that for each pair $i, j = 1, 2, \ldots, m$, $i \neq j$ there exist indices $i_0 = i, \ldots, i_s = j$ such that $T_n^{i_{k-1}} \cap T_n^{i_k} \neq \emptyset$ for $k = 1, \ldots, s$. This assumptions mean that any two stock exchanges are either "correlated" directly or through a "chain" of other stock exchanges. The latter assumption is no actual restriction as any collection of stock exchanges can be decomposed in connected subgraphs of such chains that can be analyzed independently.

Therefore, the observed price vectors $\boldsymbol{s}_t$ are elements of $(\mathbb{R} \cup \{\mathrm{NA}\})^p$ with

$$(\boldsymbol{s}_t)^j \in \begin{cases} \mathbb{R} & \exists i \in \{1, 2, \ldots, m\} \text{ such that } t \in T_n^i \text{ and } j \in I_i, \\ \{\mathrm{NA}\} & \text{otherwise} \end{cases} \tag{3.4}$$

where NA (acronym for "not available") denotes missing entries. If $t \notin \bigcup_{i=1}^m T_{\Delta t}^i$, the observation $\boldsymbol{s}_t = (\mathrm{NA}, \mathrm{NA}, \ldots, \mathrm{NA})'$ is missing entirely. Otherwise, $\boldsymbol{s}_t$ contains both numeric values and NAs. As before, we compute the log-differences

$$\boldsymbol{x}_k := \log(\boldsymbol{s}_{t_k}) - \log(\boldsymbol{s}_{t_{k-1}}) \text{ for } k = 1, 2, \ldots, n,$$

where we use the convention $\log(\mathrm{NA}) := \mathrm{NA}$, $\mathrm{NA} \pm \mathrm{NA} := \mathrm{NA}$, $x \pm \mathrm{NA} := \mathrm{NA}$ for $x \in \mathbb{R} \cup \{\mathrm{NA}\}$. Eliminating all $\boldsymbol{x}_k = (\mathrm{NA}, \mathrm{NA}, \ldots, \mathrm{NA})'$ and renumbering the observations, we obtain $n' < n$ observation vectors $\boldsymbol{x}_1, \boldsymbol{x}_2, \ldots, \boldsymbol{x}_{n'} \in (\mathbb{R} \cup \{\mathrm{NA}\})^p$ with non-missing or partially missing entries.

A straightforward naïve approach to estimating $F(\cdot)$ from $\boldsymbol{x}_1, \boldsymbol{x}_2, \ldots, \boldsymbol{x}_{n'}$ would be to filter out observations with missing entries and use the remaining observations (with all entries being real) to construct an empirical c.d.f. $\hat{F}_n(\cdot)$. Even apart from the extreme case when no such observations exist (e.g., when US, European and Asian stock exchanges are considered simultaneously), this approach is not recommendable due to a drastic decrease in statistical efficiency.

We propose an alternative approach using multiple imputation with chain equations (MICE) based on machine learning [30]. Since our scenario falls under the "missing at random" (MAR) framework of MICE due to the nature of stock data collection, we can use it to obtain n_{imp} imputed copies $\boldsymbol{x}_1^{(i)}, \boldsymbol{x}_2^{(i)}, \ldots, \boldsymbol{x}_{n'}^{(i)}$ ($i = 1, 2, \ldots, n_{\mathrm{imp}}$) of the dataset $\boldsymbol{x}_1, \boldsymbol{x}_2, \ldots, \boldsymbol{x}_{n'}$. This leads to the estimate

$$\hat{F}_n(\cdot) = \frac{1}{n_{\text{imp}} \cdot n'} \sum_{i=1}^{n_{\text{imp}}} \sum_{k=1}^{n'} \mathbb{1}_{(-\infty, \boldsymbol{x}_k^{(i)}]}(\cdot) \tag{3.5}$$

of $F(\cdot)$. Moreover, we can estimate cdf of the conditional distribution $\mathbf{E}[\boldsymbol{X} \mid \boldsymbol{X}_J = \boldsymbol{x}_k]$ for any $\boldsymbol{x}_k$ $(k = 1, 2, \ldots, n')$ with partially missing values, say, in all components $I \subset \{1, 2, \ldots, p\}$, via

$$\hat{F}_{n,k}(\cdot) \equiv \hat{F}_{\mathbf{E}[\boldsymbol{X} \mid \boldsymbol{X}_J = \boldsymbol{x}_k^J]}(\cdot) = \frac{1}{n_{\text{imp}}} \sum_{i=1}^{n_{\text{imp}}} \mathbb{1}_{(-\infty, \boldsymbol{x}_k^{(i)}]}(\cdot), \quad J := \{1, 2, \ldots, p\} \backslash I. \tag{3.6}$$

Plugging the empirical cdf from Eq. (3.5) into (3.2), the estimates $\hat{\boldsymbol{\mu}}$ and $\hat{\boldsymbol{\Sigma}}$ follow, which, in turn, produce estimates $\hat{\boldsymbol{a}}$ and $\hat{\boldsymbol{B}}$ according to Eq. (3.3). While the conventional approach to estimating pairwise covariances $\boldsymbol{\Sigma}_{ij}$ discards all observations for which either the i-th or the j-th component is missing, our approach uses multiple batches of imputed observation vectors. This not only directly ensures the estimated covariance is positive semidefinite in contrast to the conventional approach that typically produces indefinite symmetric matrices (i.e., with both positive and negative eigenvalues), but mitigates underestimation effects and boosts uncertainly quantification by creating multiple batches of imputed observations to better account for natural variability.

4   Simulation and Forecasting

With the nonparametric estimates $\hat{F}_n(\cdot)$ and $\hat{F}_{\mathbf{E}[\boldsymbol{X} \mid \boldsymbol{X}_J = \boldsymbol{x}_k^J]}(\cdot)$ at hand, we can predict the evolution of stock prices in the future. Since we are interested in risk analysis and portfolio valuation instead of option pricing, both estimation and forecasting in this paper are performed under the physical measure [28].

4.1   Simulation Under Physical Measure

To describe the procedure, we begin with the situation described in Sect. 3.1. Letting N represent the number of replications in the Monte Carlo simulation, the r-th path of the geometric random walk in Eq. (2.6) is simulated as

$$\hat{\boldsymbol{s}}_{t_k}^{(r)} = \exp(\boldsymbol{\zeta}_k^{(r)}) \hat{\boldsymbol{s}}_{t_{k-1}}^{(r)} \quad \text{for } k = n+1, n+2, \ldots, \quad \hat{\boldsymbol{s}}_{t_n}^{(r)} = \boldsymbol{s}_{t_n} \tag{4.1}$$

for $r = 1, 2, \ldots, N$, where $\boldsymbol{\zeta}_k^{(r)} \overset{\text{i.i.d.}}{\sim} \hat{F}_n(\cdot)$ for $r = 1, \ldots, N$ and $k = n+1, n+2, \ldots$. The log-increments are obtained by sampling from the empirical cdf $\hat{F}_n(\cdot)$ or, equivalently, by randomly sampling from $\boldsymbol{x}_1, \boldsymbol{x}_2, \ldots, \boldsymbol{x}_n$. Note that neither $\hat{\boldsymbol{a}}, \hat{\boldsymbol{B}}$ nor $\hat{F}_0(\cdot)$ are directly required here.

Turning to the more realistic situation of Sect. 3.2, for the i-th stock, let

$$t_{\text{last not NA}}^i := \begin{cases} \max\{t \mid t \in T_n\}, T = \{t \in T_n \mid s_t^i \neq \text{NA}\} & \text{if } T \neq \emptyset, \\ n, & \text{otherwise.} \end{cases}$$

For the sake of convenience, denote $t^i_{\text{last not NA}} := t_{k^i_{\text{last not NA}}}$ for appropriate $k^i_{\text{last not NA}} \in \{1, 2, \ldots, n\}$. For non-stop operating exchanges, $t^k_{\text{last not NA}} \equiv n$.

The r-th path is then simulated as

$$\left(\hat{s}^{(r)}_{t_k}\right)^j = \text{NA if } j \leq t^i_{\text{last not NA}}, \quad \left(\hat{s}^{(r)}_{t_k}\right)^I = \left(\exp(\zeta^{(r)}_k)\hat{s}^{(r)}_{t_{k-1}}\right)^I$$

with $I = \{i \,|\, t_k > t^i_{\text{last not NA}}\}$ for $k = \min\{k^i_{\text{last not NA}} \,|\, i = 1, 2, \ldots, p\}, \ldots, n$, where $\zeta^{(r)}_k \overset{\text{i.i.d.}}{\sim} \hat{F}_{n,k}(\cdot)$ with the latter empirical cdf defined in Eq. (3.6).

Hence, for the i-th of the p stocks, we have simulated N price curves from $t^i_{\text{last not NA}}$ to t_n. Continuing at t_n, we further obtain

$$\hat{s}^{(r)}_{t_k} = \exp(\zeta^{(r)}_k)\hat{s}^{(r)}_{t_{k-1}} \text{ for } k = n+1, n+2, \ldots, \quad \hat{s}^{(r)}_{t_n} = s_{t_n} \qquad (4.2)$$

for $r = 1, 2, \ldots, N$, where $\zeta^{(r)}_k \overset{\text{i.i.d.}}{\sim} \hat{F}_n(\cdot)$ for $r = 1, \ldots, N$ and $k = n+1, n+2, \ldots$. Using Eqs. (4.1) and (4.2), we have thus computed N paths, where the i-th path starts at $t^i_{\text{last not NA}}$ and can be continued into the future.

4.2   Forecast Bounds

Using the N simulated paths from Eq. (4.2), for any t_k, $k \geq n + 1$, and any confidence level $\alpha \in (0, 1)$, we can construct nonsimultaneous forecast intervals

$$\boldsymbol{l}_{t_k,\text{nonsim}} = \left((\hat{s}^i_{t_k})_{\alpha/2}\right)^p_{i=1}, \quad \boldsymbol{u}_{t_k,\text{nonsim}} = \left((\hat{s}^i_{t_k})_{1-\alpha/2}\right)^p_{i=1} \qquad (4.3)$$

where $(\hat{s}^i_{t_k})_\alpha$ denotes the empirical α-th quantile computed from simulated stock prices $\hat{s}^1_{t_k}, \ldots, \hat{s}^N_{t_k}$. Bonferroni simultaneous intervals can be obtained by replacing α in Eq. 4.3 with α/p. As p increases, the latter intervals become prohibitively loose limiting their practical applicability. The lower and upper bounds in Eq. (4.3) are security holder's and counterparty's VaR, respectively. Other traditional risk measures, such as CVaR, RVaR and EVaR, etc., can be computed accordingly from simulated $\hat{s}^i_{t_k}$'s. Thus, our model seamlessly integrates into existing risk management frameworks.

4.3   Portfolio Analysis

Consider a portfolio consisting of $\nu_1, \nu_2, \ldots, \nu_p$ instances of the p stocks traded at the m stock exchanges. At time t_k, $k \geq n$, the portfolio value reads as

$$V_{t_k} = \sum_{i=1}^p \nu_i S^i_{t_k}. \qquad (4.4)$$

Using the N simulated stock price vectors, the empirical push-forward measure is given via $\mathrm{d}\hat{F}_{V_{t_k}}(\cdot) = \frac{1}{N} \sum_{r=1}^N \sum_{i=1}^p \nu_i \delta_{\{\hat{s}^{(r)}_{t_k}\}}(\cdot)$ with $\delta_{\{s_0\}}(\cdot)$ denoting Dirac's delta-measure at s_0. Thus, the sample mean and sample variance of V_{t_k} read

$$\bar{V}_{t_k} = \sum_{i=1}^p \nu_i \bar{S}^i_{t_k}, \quad s^2_{V_{t_k}} \equiv \sum_{i,j=1}^p \nu_i \nu_j \widehat{\text{cov}}(S^i_{t_k}, S^j_{t_k}) \equiv \text{tr}\left[(\nu \otimes \nu) \widehat{\text{cov}}(S^i_{t_k}, S^j_{t_k})\right].$$

Likewise, the α-th quantile of V_{t_k} is estimated via respective sample quantile of $\left\{ \sum_{i=1}^{p} \nu_i \hat{s}_{t_k}^{(1)}, \sum_{i=1}^{p} \nu_i \hat{s}_{t_k}^{(2)}, \ldots, \sum_{i=1}^{p} \nu_i \hat{s}_{t_k}^{(N)} \right\}$.

5   Example

Our analysis employs a dataset comprising 52,754 observations on 324 stocks, recorded at one-minute intervals, spanning from April 22, 2019, at 13:30:00 UTC, to July 26, 2019, at 19:59:00 UTC. The dataset was sourced from Dukascopy's historical market data archives [10]. Given the global scope of our study, data from two major stock exchanges – New York Stock Exchange (NYSE) and London Stock Exchange (LSE) – located in two distinct time zones, were integrated. This inclusion necessitated addressing instances of missing observations attributable to differing operational hours. Historical data from early 2019 were chosen since the market was not overly turbulent during those times.

For the purposes of this study, the dataset was divided into two segments: the first one, encompassing 70.94% of the observations, spans from the start date up until June 28, 2019, at 19:59:00 UTC. This segment was used for model calibration. The subsequent segment, containing the remaining 29.06% of the data, covers the period from July 1, 2019, at 07:00:00 UTC, to the end of the dataset. This latter portion was utilized for out-of-sample testing and benchmarking.

5.1   Exploratory Data Analysis

Some essential exploratory data analysis (EDA) was performed on the estimated mean vector and covariance matrix of the original (unimputed) log-increments. Let $\hat{\mu}$ and $\hat{\Sigma}$ denote the estimated mean vector and covariance matrix using the technique described in [17]. This approach is methodologically superior to the conventional procedure based on pairwise complete observations. Multiplying the empirical drift vector $\hat{\mu}$ with $252 \times 24 \times 60$ to obtain annual returns (note that $\hat{\mu}$ is based on 1-minute returns and there are 252 trading days in a year), the distribution of annual returns for the 246 US and 78 UK stocks is displayed in Fig. 1. While UK stocks tended to produce better returns in the pre-pandemic year 2019, the difference was not sufficiently pronounced to separate the markets.

To better visualize the two stock markets, we use the well-known t-SNE [31] procedure (`tsne()` routine with default `perplexity` $= 30.0$ from the eponymous R package) was applied to embed the data into a 2D space based on the estimated mean squared distance between log-increments ζ_i and ζ_j for each pair of stock:

$$\widehat{\mathbb{E}}\left[(\zeta_i - \zeta_j)^2\right] = \left(\widehat{\mathbb{E}}[\zeta_i] - \widehat{\mathbb{E}}[\zeta_j]\right)^2 + \widehat{\mathrm{Var}}[\zeta_i] + \widehat{\mathrm{Var}}[\zeta_j] - 2\,\widehat{\mathrm{Cov}}[\zeta_i, \zeta_j],$$

where respective variances/covariances were obtained from $\hat{\Sigma}$. The results are displayed in Fig. 2. The US stocks were observed to be well-separated from their UK counterparts suggesting within-market correlations are stronger than between-market correlations. These separation effects could have potentially

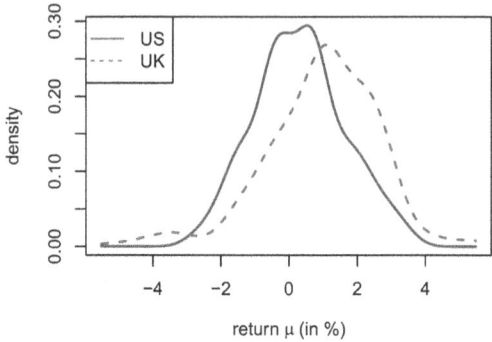

Fig. 1. US and UK annualized stock returns

been magnified by "deflated" between-market correlations in stock price log-increments resulting from missing data due to partially overlapping trading times. Though appropriate techniques aimed to minimize these effects were implemented in this paper [17], further investigations based on after-hours trading data will need to be thoroughly investigated to make more definitive conclusions.

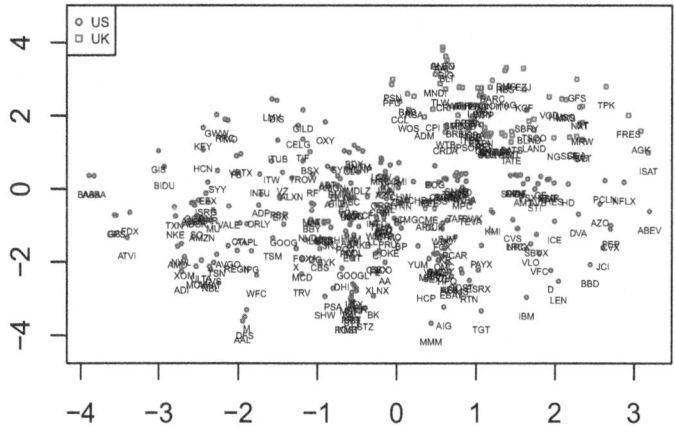

Fig. 2. 2D embedding for US and UK annual stock returns

5.2 Backtesting and Benchmarking

Next, the historical log-differences in the training set (from April 22, 2019, 13:30:00 UTC to July 1, 2019, 07:00:00 UTC) were imputed as described in Sect. 3.2 using `mice()` routine from the eponymous R package with $m = 5$ (viz. n_{imp} in Eq. (3.5)), `maxit = 10` and `method = rf` (random forest imputation). For each of 324 stocks, we analyzed the stock price, log-returns, marginal cdf and probability density function (pdf), autocorrelation function (acf) and partial autocorrelation function (pacf). See Fig. 3 for Amazon Corp stock (ticker AMZN). A full set of 324 plots is provided in Supplement. Most stock analyzed exhibited some degree of non-Gaussianity. The amount of autocorrelation was modest suggesting our proposed semiparametric model is adequate.

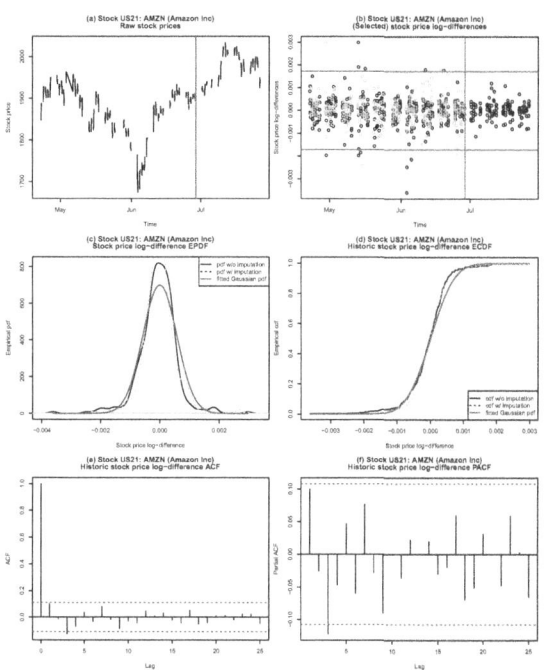

Fig. 3. Analysis of three months of historical training data for `AMZN` stock

Having calibrated our semiparametric geometric random walk model in Eq. (2.6) and estimated the mean vector and the covariance matrix from the training data, we performed a set of backtesting and benchmarking analyses based on one month's worth of stock data starting on July 1, 2019, 09:31:00 UTC (opening time of LSE). For backtesting purposes, we first simulated 5,000 paths using the procedure described in Sect. 4.1. For each of the 324 stocks, we constructed non-simultaneous 95% forecast regions. For example, Fig. 4 displays historic (before July 1, 2019) and backtesting (after July 1, 2019) 1-minute AMZN stock prices where "gaps" in the graph correspond to market closures on non-trading days (weekends and holidays). The solid blue region encloses 95% pointwise forecast bounds for the AMZN stock throughout July, 2019, where the lower and upper bounds are 2.5% VaR for the stock holder and the counterparty, respectively. The forecast region can be interpreted as the expected range, within which the stock price is likely to fluctuate with a 95% confidence at any given time. The full set of 324 plots is provided in Supplement.

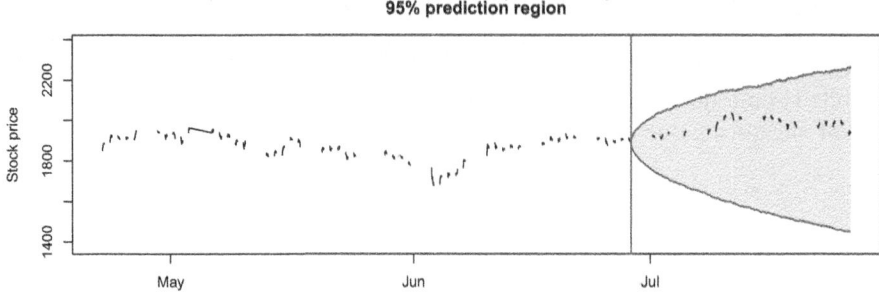

Fig. 4. 95% pointwise forecast bounds for AMZN stock for July 2019

Next, we performed a benchmarking study to compare our semiparametric model to Gaussian GBM as well as alternative geometric random walks with multivariate Laplace and Student's t-like log-increments. Calibrating the latter three parametric models using the method of moments based on the mean vector and covariance matrix estimated according to [17] with appropriate spectral augmentations to assure positive definiteness and a condition number of below 50, Eq. (4.2) with Gaussian, Laplace and Student's t-like innovations $\zeta_k^{(r)}$ was used to simulate stock prices. The effective "dimension" and degrees of freedom for Student's t log-increments were estimated to be $p' = 106$ and $\nu = 10.48$, respectively, based on optimal Kolmogorov-Smirnov distance of 0.056.

Using simulated stock prices, we considered a hypothetical portfolio of a nominal $1 worth on July 1, 2019, 09:31:00 UTC consisting of top five US and top five UK stocks based on their share in S&P 500 [33] and iShares MSCI United Kingdom ETF [32] as of March 6, 2024. (Historical holdings data for July 1, 2019 were not available.) Table 1 displays the stock weights with US and UK stocks making 80% and 20% of the total portfolio, respectively.

Table 1. Stock holdings in the portfolio

US ticker	MSFT	AAPL	NVDA	AMZN	GOOGL
Weight	24.262%	20.655%	15.993%	12.726%	6.363%
UK ticker	AZN	HSBA	ULVR	DGE	REL
Weight	6.243%	4.677%	3.874%	2.626%	2.580%

We generated $N = 500$ independent paths our semiparametric and the three competing parametric (Gaussian, Student's t and Laplace) models. At each grid time t in July 2019, the portfolio value was computed using Eq. (4.4) with the stock shares $\nu_i = w_i/S_{t_0}^i$, where w_i's are respective holdings in Table 1 and

$S_{t_0}^i$'s are (known) stock prices at the beginning of the backtesting/benchmarking period (July 1, 2019, 09:31:00 UTC).

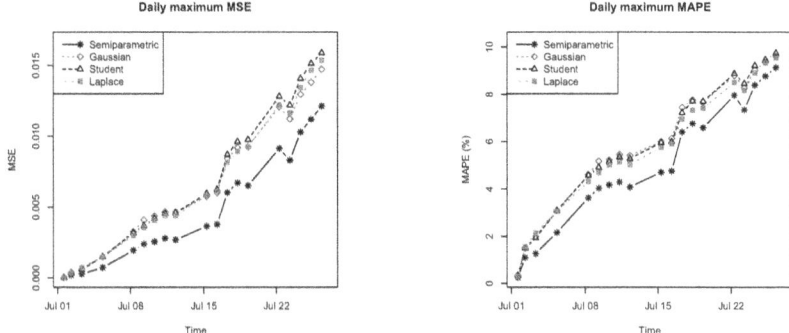

Fig. 5. Estimated daily maximum errors for July 2019

The usual empirical mean squared error (MSE) and mean absolute percentage error (MAPE) were used as error measures:

$$\widehat{\text{MSE}}(t) = \frac{1}{N} \sum_{r=1}^{N} \left(\hat{V}_t^{(r)} - V_t^{\text{obs}} \right)^2, \quad \widehat{\text{MAPE}}(t) = \frac{100\%}{N} \sum_{r=1}^{N} \frac{\left| \hat{V}_t^{(r)} - V_t^{\text{obs}}(t) \right|}{V_t^{\text{obs}}} \quad (5.1)$$

where V_t^{obs} is the observed portfolio value and $\hat{V}_t^{(r)}$ is the r-th simulated price at time t. As can be seen from Fig. 5, our approach outperforms all three competing parametric models at forecasting the portfolio value. The conclusions also agree with observations presented in [22].

6   Summary

In this paper, we have proposed a new stock market model that leverages a multivariate geometric random walk without imposing any parametric assumption or structural assumption on the log-increments. Our model offers greater flexibility in representing return distributions and creates a significant link with the continuous-time Geometric Brownian Motion, bridging discrete and continuous modeling methodologies. With its efficacy demonstrated through application to stocks across major exchanges and subsequent rigorous bechmarking and backtesting, our model not only enhances predictive accuracy but also provides a solid foundation for risk management practices. By addressing extreme market events and anomalies, it represents a substantial shift towards a more comprehensive and empirically grounded approach in financial modeling and risk assessment.

Despite our primary focus on uncertainty quantification on the tails, we recognize the growing significance of nonlinear autoregressive models, such as ANN and RF, in capturing complex market dynamics [2,3,20]. While these and other machine learning methods have demonstrated substantial promise in financial forecasting, flexible, robust and computationally efficient algorithms that are capable of analyzing high-dimensional high-frequency data with both missing values and heavy tails are yet to be developed. As part of our future research, a combination of the semiparametric approach developed in this paper with recent developments in autoregressive deep learning for time series [11] and other machine learning techniques has the potential to provide new promising pathways to better predictive capabilities, uncertainty quantification and practical applicability in risk management and other arenas.

Supplemental Files

Supplemental files are available at https://github.com/mpokojovy/Semipar GBM.

References

1. Abad, P., Benito, S., López, C.: A comprehensive review of value at risk methodologies. Span. Rev. Financ. Econ. **12**(1), 15–32 (2014)
2. Ahmed, A., Khalid, M.: Multi-step ahead wind forecasting using nonlinear autoregressive neural networks. Energy Procedia **134**, 192–204 (2017)
3. Benrhmach, G., Namir, K., Namir, A., Bouyaghroumni, J.: Nonlinear autoregressive neural network and extended Kalman filters for prediction of financial time series. J. Appl. Math. **1–6**, 2020 (2020)
4. Bouchaud, J.P., Potters, M.: Theory of Financial Risk and Derivative Pricing: from Statistical Physics to Risk Management. Cambridge University Press, Cambridge (2003)
5. Cator, E.A., Lopuhaä, H.P.: Central limit theorem and influence function for the MCD estimators at general multivariate distributions. Bernoulli **18**(2), 520–551 (2012)
6. Che, Z., Purushotham, S., Cho, K., Sontag, D., Liu, Y.: Recurrent neural networks for multivariate time series with missing values. Sci. Rep. **8**(1), 6085 (2018)
7. Chen, L., Pelger, M., Zhu, J.: Deep learning in asset pricing. Manage. Sci. **70**(2), 714–750 (2021)
8. Cho, K., et al.: Learning phrase representations using RNN encoder-decoder for statistical machine translation. In: Proceedings of the 2014 Conference on Empirical Methods in Natural Language Processing, EMNLP, pp. 1724–1734 (2014)
9. de Miranda Cardoso, J.V., Ying, J., Palomar, D.: Graphical models in heavy-tailed markets. In: Advances in Neural Information Processing Systems, vol. 34, pp. 19989–20001 (2021)
10. Dukascopy Bank SA. Historical data feed (2019). https://www.dukascopy.com/swiss/english/marketwatch/historical/. Accessed 15 Mar 2024
11. Fouladgar, N., Främling, K.: A novel LSTM for multivariate time series with massive missingness. Sensors **20**(10), 2832 (2020)
12. Gregory, J.: The xVA Challenge: Counterparty Credit Risk, Funding, Collateral and Capital. The Wiley Finance Series, 3rd edn. Wiley, Chichester (2015)
13. Gu, S., Kelly, B., Xiu, D.: Empirical asset pricing via machine learning. Rev. Financ. Stud. **33**(5), 2223–2273 (2020)

14. Hochreiter, S., Schmidhuber, J.: Long short-term memory. Neural Comput. **9**(8), 1735–1780 (1997)
15. Jorion, P.: Value at Risk: The New Benchmark for Managing Financial Risk. McGraw-Hill, New York (2007)
16. Kalkbrener, M., Packham, N.: Stress testing of credit portfolios in light-and heavy-tailed models. J. Risk Manag. Financ. Inst. **8**(1), 34–44 (2015)
17. Lounici, K.: High-dimensional covariance matrix estimation with missing observations. Bernoulli **20**(3), 1029–1058 (2014)
18. Omelchenko, V.: Parameter estimation of geometric stable distributions. Kybernetika - Praha **50**(6), 1–21 (2014)
19. Platen, E., Rendek, R.: Exact scenario simulation for selected multi-dimensional stochastic processes. Commun. Stoch. Anal. **3**(3), 443–465 (2009)
20. Polyzos, E., Siriopoulos, C.: Autoregressive random forests: machine learning and lag selection for financial research. Comput. Econ. (2023)
21. Prakash, A., James, N., Menzies, M., Francis, G.: Structural clustering of volatility regimes for dynamic trading strategies. Appl. Math. Finan. **28**(3), 236–274 (2021)
22. Pun, C.S., Wong, H.Y.: Resolution of degeneracy in Merton's portfolio problem. SIAM J. Financ. Math. **7**(1), 786–811 (2016)
23. Reddy, K., Clinton, V.: Simulating stock prices using geometric Brownian motion: evidence from Australian companies. Australas. Account. Bus. Finan. J. **10**(3), 23–47 (2016)
24. Rockafellar, R.T., Uryasev, S.: Conditional value-at-risk for general loss distributions. J. Bank. Finan. **26**(7), 1443–1471 (2002)
25. Ruiz, I.: XVA Desks - A New Era for Risk Management: Understanding, Building and Managing Counterparty, Funding and Capital Risk. Applied Quantitative Finance. Palgrave Macmillan UK, London (2015)
26. Schmitt, N.: Heterogeneous expectations and asset price dynamics. Macroecon. Dyn. **25**(6), 1538–1568 (2021)
27. Segal, M., Xiao, Y.: Multivariate random forests. Wiley Interdisc. Rev. Data Mining Knowl. Discov. **1**(1), 80–87 (2011)
28. Strommen, S.J.: Understanding the Connection Between Real-World and Risk-Neutral Scenario Generators. Society of Actuaries Research Institute, Schaumberg (2022)
29. Tang, F., Ishwaran, H.: Random forest missing data algorithms. Stat. Anal. Data Mining ASA Data Sci. J. **10**(6), 363–377 (2017)
30. Van Buuren, S.: Flexible Imputation of Missing Data. Interdisciplinary Statistics Series, 2nd edn. Chapman & Hall/CRC, Boca Raton (2018)
31. Van der Maaten, L., Hinton, G.: Visualizing data using t-SNE. J. Mach. Learn. Res. **9**(11) (2008)
32. VettaFi. iShares MSCI United Kingdom ETF (2024). https://etfdb.com/etf/EWU/#holdings
33. VettaFi. S&P500 ETF (2024). https://etfdb.com/etf/SPLG/#holdings

Disease Similarity and Disease Clustering

Drew Brady[✉] and Hisham Al-Mubaid

University of Houston – Clear Lake, Houston, TX, USA
{Bradyd6193,Hisham}@uhcl.edu

Abstract. In bioinformatics research, identifying the common characteristics of human diseases is crucial for understanding disease-disease associations. In this direction, disease similarity and disease clustering methodologies provide many insights by identifying underlying patterns, disease mechanisms, and other factors associated with disease causation and progression. Furthermore, the information and outcomes from these insights can inform drug repurposing, precision medicine, and other medical and healthcare applications. However, there are limitations in computational techniques, data integration, and disease knowledge that prevent or hinder a comprehensive understanding of the relationships between diseases. With our study, we aim to assist ongoing research by providing an integrated overview of disease similarity and disease clustering. We will examine the biomedical databases, disease databases and vocabularies, and disease-related information commonly used for this research. Then, we analyze some of the computational methods and integrated approaches used to quantify disease-disease associations. Additionally, we provide insights that can be derived from disease similarity and clustering analysis.

Keywords: Disease Similarity · Disease Clustering · Human Disease Analysis

1 Introduction

Most of our knowledge and information about various human diseases are organized in structured databases known as knowledge organization systems (KOSs) [1–3]. These systems use terms or classifications to define and represent diseases and their complex relationships with biological and clinical features. Ontologies, classification systems, and vocabularies are just a few categories of concept systems encompassed in KOSs that allow for the representation, organization, and analysis of diseases and disease-related data [1–3]. Furthermore, the information provided by these systems is broadly categorized as disease vocabularies, disease annotations, and gene functional annotations [3, 4]. Disease vocabulary is a broad term for the databases that provide classifications and definitions for diseases and disease-related entities [1–4]. Disease annotations provide detailed information on the relationships between disease terms and disease-related data, such as disease-phenotype, disease-gene, and disease-SNP associations [3, 4]. Gene functional annotations describe the functional properties of genes and gene products within the molecular, cellular, and biological domains [3, 4].

H. Han (Ed.): SDSC 2024, CCIS 2158, pp. 64–77, 2024.
https://doi.org/10.1007/978-3-031-67871-4_5

In disease similarity and clustering research, KOSs are crucial for identifying and understanding the associations between diseases. With disease similarity, we aim to measure how closely related two diseases are based on shared characteristics [3–5]. Various methods, such as pairwise metrics, text-mining, and networks, are used to quantify the similarity between two diseases. For disease clustering, clustering algorithms are used to identify groups of closely related diseases based on shared characteristics [5, 6]. The commonly used categories of clustering algorithms include hierarchical, centroid-based, distribution-based, density-based, and deep learning-based clustering.

This study investigates the various aspects of disease similarity and disease clustering. We examine the various types of disease-related data utilized in this area of research, and we discuss the concept systems that organize and classify this information. Moreover, we examine the computational methods used to measure the similarities between diseases and provide insights into their limitations. Our review aims to encourage future research that will enhance our understanding of disease-disease associations.

2 Background and Literature Review

Disease similarity and disease clustering have significant roles in investigating the complex nature of diseases [3–5]. Specifically, disease similarity and clustering analyses have revealed that certain diseases may share commonalities that can provide insights that may inform disease diagnosis, prevention, and treatment [3–6]. The concepts of disease similarity and disease clustering both aim to elucidate disease similarities and relationships. However, they differ in the focus of these relationships and the methods used for research.

Research in disease similarity measures the semantic similarity or relatedness between two diseases, often quantified with similarity scores using pairwise, path-based, and node-based methods. Semantic similarity is a specific form of semantic relatedness that refers to the degree of similarity between two diseases within a structured framework such as an ontology [3, 4]. In contrast, semantic relatedness encompasses the broader scope of information overlap between diseases, such as phenotypes, genes, or biological processes [3, 4]. Su et al. (2019) demonstrated an application of semantic similarity through the development of GPSim [7]. Their approach measures the semantic similarity between diseases by evaluating the extracted ontological associations between diseases, genes, and phenotypes. They compared the results from GPSim to other established methods using the evaluation metrics ROC (Receiver Operator Characteristic) and AUC (Area Under Curve).

Further advancements in computational methods for disease similarity include text-mining, networks, and graph neural networks. For example, Rosario-Ferreira et al. (2021) illustrated an advanced application of text-mining with SicknessMiner, a deep learning text-mining tool for extracting disease-disease associations in biomedical literature [8]. In another study, Li et al. (2021) employed deep learning representation on a gene network to measure disease similarity [9]. They also used ROC and AUC as evaluation metrics and compared the results to established semantic similarity measures.

Research in disease clustering utilizes unsupervised machine-learning tasks called clustering algorithms to identify meaningful categories of diseases [5, 6]. These methods typically analyze feature-vector representations of diseases, which are measurable characteristics of diseases [3, 5]. Brady and Al-Mubaid (2024) utilized a centroid-based clustering algorithm, K-means, to cluster disease feature-vector representations using biological process (bp) annotations [10]. They assessed the cluster quality through a two-step process. Initially, they used edge-based approaches to evaluate the semantic relationships within the Disease Ontology (DO) and then validated the results with the disease similarity results from Yang et al. (2023) [6, 12].

The cluster analysis performed by Yang et al. (2023) integrated gene expressions, biological processes, and phenotypes of diseases in a multi-view human disease network (MVHDN). They initially produced three single-view networks for each disease-related data type and then applied a similarity network fusion (SNF) model to combine the three networks. Then, they utilized a spectral clustering algorithm to classify clusters from the MVHDN. In another study, Sherif and Ahmed (2022) present a deep learning-based clustering method that uses a deep convolutional autoencoder to identify genetically similar virus strains of SARS-CoV-2 [11]. Their method outperformed traditional clustering approaches such as K-means and hierarchical clustering by identifying patterns in complex, noisy biological data sets.

3 Knowledge Organization Systems

The systematic organization and classification of information are crucial for ensuring accessibility, consistency, and knowledge discovery in any research field [1–3]. In bioinformatics, knowledge organization systems (KOS) are pivotal in achieving this objective for disease similarity and clustering research [1–3]. Ontologies, classification systems, and other similar concept systems are tools for organizing information but differ in structure and purpose [1]. Ontologies are hierarchical systems that enable the standardization of terms and relationships (e.g., *is_a, part_of,* and *contributes_to*) within a specific domain [2, 3]. Figure 1 illustrates a small part of the hierarchical structures, biological entities, and their relationships within the Gene Ontology (GO), visualized using a web-based tool called *QuickGO* [13–15]. An ontology provides a framework for data integration, interoperability, and analysis of disease-related data [1–3]. Classification systems are like ontologies; they may utilize hierarchical arrangements to group similar terms. However, they are considered less semantically rich due to their lack of scope and representation of relationships between terms [1]. Controlled vocabularies are another type of KOS that standardizes the terminology and language within a specific domain [1]. They are often integral components of ontologies used to represent the concepts describing objects and their relations. For this paper, we may use the terms *ontology* and *vocabulary* interchangeably.

The Disease Ontology (DO) and Medical Subject Headings (MeSH) are prime examples of what ontologies can provide [12, 16]. DO provides a controlled vocabulary for classifying information about diseases and their hierarchical relationships. It organizes disease terms with hierarchical structures to represent semantic relationships (parent-child associations), denoted by the *is-a* qualifier. As shown in Fig. 2, DO classifies angiosarcoma (DOID:0001816) as a subtype, or a child, of vascular cancer (DOID:175) [12].

Furthermore, DO integrates a wide range of disease-related data, such as genetic features, clinical features, and environmental factors, for additional resources they provide.

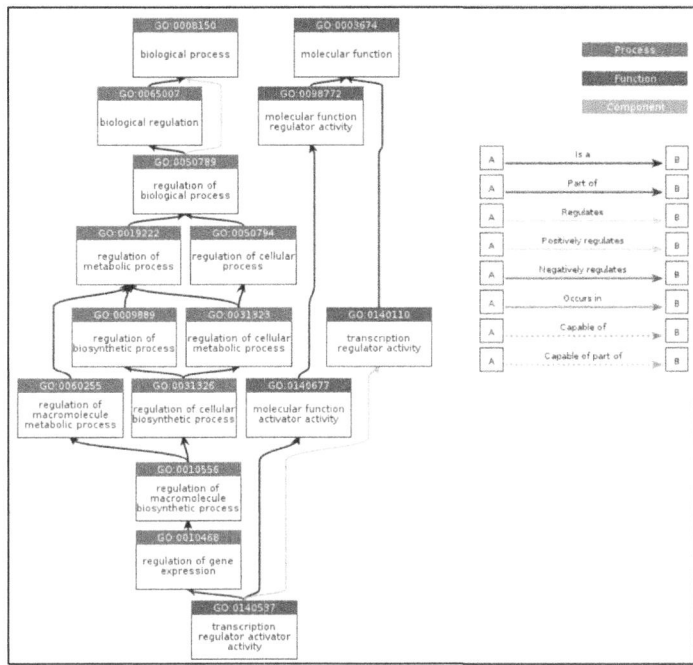

Fig. 1. An ancestor chart from QuickGO that shows the hierarchical arrangement and ontology relations between biological process and molecular function GO terms from the Gene Ontology (GO).

It also enriches its resources by integrating cross-references from other disease vocabularies, such as *MeSH*, *OMIM*, and *HPO* [16–18]. MeSH is a standardized vocabulary of biomedical information created by the National Library of Medicine (NLM). Like DO, MeSH utilizes tree structures to hierarchically categorize terms into several categories, including but not limited to diseases, drugs and chemicals, anatomy, and organisms.

While not an ontology, the Online Mendelian Inheritance of Man (OMIM) is another widely used resource for disease-related information, namely genes and phenotypes [17]. The phenotypic entries in OMIM provide information on clinical features and management, inheritance, and diagnosis, while the gene entries include information such as gene mappings, structures, expressions, and functions. OMIM extensively catalogs the relationships between genes and phenotypes by including their associations for each entry. For example, OMIM provides and maintains the *morbidmap* dataset that contains more than 5000 gene-phenotype relationships and details the cytogenetic location for mapped human diseases. Besides these resources, other KOSs like the Gene Ontology (GO) and Human Phenotype Ontology (HPO) are also fundamental for this research. GO is a controlled vocabulary that organizes gene-related information into three major aspects [13–15]. The Gene Ontology Annotations (GOA) database uses GO to connect genes and gene products to their molecular functions, cellular components, and biological processes [15]. HPO is another widely used ontology that provides information for phenotypic abnormalities associated with human diseases [18].

While these KOSs are crucial for disease similarity and clustering analyses, they have limitations like incompleteness and inconsistency that impact the accuracy and reliability of such analyses [2, 3, 6, 19, 20]. The incompleteness of terms often results from inadequate human genome coverage and leads to critical gaps in our understanding of disease-biological entity relationships, namely disease-gene associations [3, 6]. Moreover, the inconsistency of disease classifications across databases hinders the interoperability of information, thus limiting comparative and integrative analyses [2, 3, 19, 20]. For instance, Zhu et al. (2020) address these challenges by identifying phenotypically similar diseases across three disease-related databases by comparing disease mappings and additionally using clinical similarities derived from disease classifications [19]. They manually evaluated 392 disease pairs and confirmed that 80% were phenotypically similar, 18% were false negatives, and approximately 4% were false positives. This study illustrates the need for accurate disease mappings and expert validation to improve disease similarity analyses. To reduce these limitations, Henry et al. (2021) present Disease Map Ontology (DMO), which is an advanced approach to constructing ontology-based models by converting disease maps into ontologies [20]. The results from their study show promise for improving ontology limitations with accuracy, consistency, and interoperability.

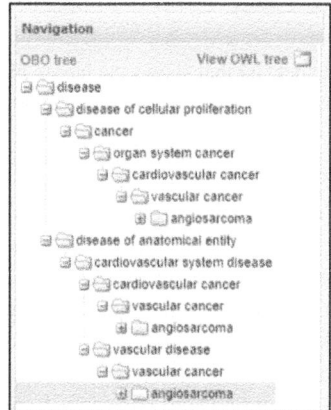

Fig. 2. An image of the OBO tree (Open Biological and Biomedical Ontology) of angiosarcoma from the Disease Ontology (DO).

4   Disease Data, Information, and Attributes

With the dozens of databases available, there is an abundance of disease-related data that can be utilized for disease similarity and disease clustering research [2–4, 21]. Cheng et al. (2019) organize this data and information into three categories: disease vocabularies, disease annotations, and gene functional annotations [4].

Disease vocabularies are standardized terminologies for classifying diseases and disease-related attributes. These resources are utilized for various applications, including disease annotations, validation techniques, semantic relatedness, and similarity measures. Hierarchical vocabularies are commonly used to determine disease similarities through semantic similarity measures, including path-based and node-based metrics [2–4].

Disease annotations refer to a database or a technique that establishes connections between a disease term from a specified resource with certain biological or clinical features [3, 4, 21]. There are many types of associations used for disease annotations. A commonly used disease association is disease-phenotype annotations, which detail observable traits, symptoms, and characteristics of a specific disease. Disease-gene annotations are another important and frequently used resource in disease similarity and clustering research. The onset or progression of a disease can be associated with mutation or alteration of the genes included in these annotations, such as protein-coding genes (PCG) [3, 21]. Disease-nRNA annotations associate diseases with non-coding RNA, such as microRNAs (miRNAs) and long non-coding RNAs (lncRNAs). The alteration of nRNA can cause dysregulation in gene expressions, which may contribute to disease mechanisms [3, 4, 21]. Disease-SNP annotations are another commonly used annotation for disease similarity research [3, 21]. Single nucleotide polymorphisms (SNPs) are a commonly occurring genomic variant that can have a range of effects on genetic function and expression [3]. Aside from the named annotations, many other disease associations used for disease similarity and clustering research include disease-drug/chemical, disease-metabolite, disease-pathway, and disease-microbe annotations [3, 21].

Gene functional annotations detail the functional aspects of genes and gene products in molecular functions, cellular components, and biological processes [3, 4, 13–15]. The molecular functions (mf) refer to the activities of gene products at a molecular level, the cellular components (cc) detail the location where the gene products are active, and the biological processes (bp) describe the biological objectives that the gene products contribute to [13–15]. The annotations of molecular functions and biological processes can be valuable resources in disease similarity and clustering research [3, 4, 6, 10]. The biological process annotations of diseases via their genes have proven effective for disease similarity and clustering research, especially when disease-gene associations are unknown or when two diseases lack shared genes directly [3, 6].

5   Computational Methods

5.1   Disease Similarity Methods

Similarity-based methods are a broad category of techniques utilized in disease similarity research [3, 4, 22–25]. These methods calculate a similarity or pairwise score to determine how closely related two diseases are, and numerous pairwise metrics can be utilized to compute these scores. These metrics are also significant for distance-based clustering algorithms, such as hierarchical and K-means clustering [23]. These pairwise metrics facilitate the comparison of two diseases, D_1 and D_2, represented by their respective disease-related entities, $d_{1,i}$ and $d_{2,i}$. As shown in Eq. 1, a disease D_q can be characterized by its features $d_{q,i}$:

$$D_q = \{d_{q,i}\}_{i=1..n} \tag{1}$$

Euclidian distance measures the distance between two disease vectors, D_1 and D_2, in an n-dimensional space, as shown in Eq. 2. This method is useful when comparing diseases based on numerical features like biological processes or genes. However, it suffers in high-dimensional spaces (curse of dimensionality) and is sensitive to varying scales across different features [23].

$$d(D_1, D_2) = \sqrt{\sum_{i=1}^{n} (d_{1,i} - d_{2,i})^2} \tag{2}$$

Cosine similarity calculates the cosine of the angle between two disease vectors, D_1 and D_2, as shown in Eq. 3. This method is preferred when the magnitude of vectors is not crucial, which makes it suitable for comparing disease vectors where only the direction of the data matters. However, this scale invariance may be a limitation when the magnitude carries significant information, which could lead to overlooking critical biological functions or signals [23].

$$sim(D_1, D_2) = cos(\theta) = \frac{\sum_{i=1}^{n}(d_{1,i} \times d_{2,i})}{\sqrt{\sum_{i=1}^{n}(d_{1,i})^2 \times (d_{2,i})^2}} \tag{3}$$

Jaccard index measures the similarity between two disease sets, D_1 and D_2, by dividing the size of their intersection by size of their union, as shown in Eq. 4. This method is ideal for comparing diseases based on binary features such as the presence or absence of genes, but it does not consider quantitative aspects like gene expression levels [23].

$$J(D_1, D_2) = \frac{|G_{D1} \cap G_{D2}|}{|G_{D1} \cup G_{D2}|} \tag{4}$$

Pearson correlation coefficient measures the linear correlation between two variables, as shown in Eq. 5. In other words, this method quantifies the direction and degree of linear association between two disease vectors, D_1 and D_2. This method assumes a linear relationship between features, which can be limiting for biological data where the relationships might be non-linear. Additionally, it cannot establish whether the changes in certain factors directly lead to changes in the degree of similarity between diseases.

$$sim(D_1, D_2) = \frac{\sum_{i=1}^{n}(d_{1,i} - \overline{d_1})(d_{2,i} - \overline{d_2})}{\sqrt{\sum_{i=1}^{n}(d_{1,i} - \overline{d_1})^2}\sqrt{\sum_{i=1}^{n}(d_{2,i} - \overline{d_2})^2}} \tag{5}$$

Semantic similarity-based methods specifically focus on measuring the hierarchical relationships within ontologies [3, 22, 24–26]. These methods typically use edge-based and node-based approaches to quantify how closely related two diseases are within a hierarchical structure [22, 24–26]. Edge-based approaches compare disease terms utilizing the structure paths or edges that connect terms in an ontology, such as DO and MeSH [12, 16]. A common metric is the shortest path length, where the semantic similarity is inversely related to the shortest path length of two terms. This method assumes that the closer the terms within an ontology reflect higher similarity. However, this assumption may not always be valid because the path length may not consistently convey the same degree of similarity in different levels of the ontology [22, 24–26].

Node-based approaches assess similarity based on the characteristics or properties of disease terms. These approaches commonly use the information content (IC) of ontological terms. The IC of a term describes that the specificity of a term is inversely correlated to the frequency of its occurrence. This means that diseases that appear less in a dataset are more informative, while those that appear often are less informative [3, 22–26]. A limitation of node-based approaches to consider is that they may not effectively elucidate the relationships between terms at differing levels of specificity [22, 24, 26]. They also rely on annotation density, which may introduce biases in semantic relationships for terms studied less [22, 24, 25].

Some commonly used node-based approaches that utilize IC are Resnik's, Lin's, and Jiang-Conrath's similarity [27–29]. Resnik's similarity (RES) quantifies the similarity between two terms based on the IC of their most informative common ancestor (MICA), as shown in Eq. 6 [27]. The limitation of this approach is that it only considers the MICA and disregards the individual contributions of the compared entities, which could result in overlooking unique attributes that are also informative [24, 25].

$$sim_{RES}(D_1, D_2) = IC(MICA(D_1, D_2)) \tag{6}$$

Lin's similarity (LIN) extends the concept of RES by considering the individual IC of each term and the IC of the MICA, as shown in Eq. 7 [28]. A notable limitation of Lin's method is the influence of the relative differences in the IC between the terms and their MICA, meaning that the measure may be affected by the MICA's position within the ontology [25, 26].

$$sim_{LINS}(D_1, D_2) = \frac{2 \times IC(MICA(D_1, D_2))}{IC(D_1) + IC(D_2)} \tag{7}$$

Jiang-Conrath similarity (JCN) also considers the IC of the MICA and the IC of each term, as shown in Eq. 8 [29]. Like Lin's method, this approach is influenced by the relative IC differences between the terms and their MICA, which may result in misleading evaluations [25, 26].

$$sim_{JCN}(D_1, D_2) = 1 - (IC(D_1) + IC(D_2) - 2 \times IC(MICA(D_1, D_2))) \tag{8}$$

Text-Mining. The data from biomedical and clinical literature has proven to be a rich information resource for disease similarity analysis [3, 30]. Information Extraction (IE) is a fundamental procedure of the text-mining process that extracts meaningful information from unstructured data [3, 30]. The Named Entity Recognition (NER) subtask of IE identifies and classifies all instances of specific categories of terms from textual data. With these identified terms, the Relationship Extraction (RE) subtask of IE identifies the implied associations between extracted terms.

The structured data obtained from IE, NER, and RE can be used in various applications, including knowledge discovery and hypothesis generation, feature-vector representations, pairwise metrics, and network analysis [3, 30]. Knowledge discovery and hypothesis generation are advanced subtasks of the text-mining process that derive hidden insights to formulate scientific hypotheses [30]. For instance, the SicknessMiner tool presented by Rosario-Ferreira et al. (2021) identified disease-disease associations from biomedical literature without requiring prior disease-gene knowledge [8].

Feature-Vector Representations. Feature-based approaches utilize vectors of disease-related characteristics to represent and measure disease similarity, as shown in Eq. 1. Studies can utilize and integrate various techniques such as pairwise metrics, text-mining, networks, and clustering by representing diseases as feature vectors. Zhang and Zang (2022) demonstrate an innovative use of feature-based approaches through their study with ImpAESim [31]. This algorithm integrates heterogenous disease-related data to produce a multi-view network, then utilizes an autoencoder model to identify informative disease feature-vector representations from the produced network. Disease similarity is subsequently evaluated by computing the cosine similarity of these representations. Their approach improves the accuracy of disease similarity results by mitigating the limitations of high-dimensional data.

Networks. Network-based approaches have become fundamental techniques for disease similarity research by utilizing graphical models to represent the complex relationships among biological entities [32]. Networks can be broadly categorized into homogeneous and heterogeneous networks. Homogeneous or single-view networks consist of a singular type of node and edge, meaning they focus on specific interactions, such as disease-gene, disease-SNP, and disease-drug associations. However, these networks may provide limited insight when representing the complexity of diseases [32]. To overcome this limitation, heterogeneous or multi-view networks integrate multiple data types to facilitate a more comprehensive analysis of disease associations.

Networks can model the functions and associations of any biological entity [32]. Some common representations for disease similarity include molecular mechanisms, text-mined data, and disease similarity scores. Analysis of these networks uses techniques including network topology, network centrality, and network motifs [32]. Each of these techniques provides insight into a network's structural and functional characteristics. Topology analysis is fundamental for understanding the overall structure of a network. A key aspect of topology is the degree distribution, which describes how edges distribute among nodes [32]. The degree distribution can denote the presence of hub nodes, which indicates a point for many pathways and interactions among other nodes [32]. The clustering coefficient is another topological feature that reveals a node's tendency to form dense communities; this can reveal how diseases share pathways [32]. Degree centrality is a measure that identifies the most important nodes (i.e., hubs) within a network. Finally, network motifs are recurring subgraphs within a network, and they can provide insights into the functional architecture of HDNs. Aside from the previously named, many more measures are utilized for network analysis.

Graph Neural Networks. In recent years, deep learning has advanced machine learning techniques with neural networks, especially with Graph Neural Networks (GNNs).

GNNs are powerful tools for identifying complex and non-linear relationships within graphical data [33, 34]. However, they can be computationally intensive, and utilizing too many layers may result in node representations becoming indistinguishable (i.e., over-smoothing). In [34], Gao et al. (2022) demonstrate the potential of GNNs in disease similarity by using a Multi-Task GNN. The results of their model outperformed the results of other methods, such as Resnik's, Lin's, and Wang's. Additionally, their approach diminishes the limitation of unknown disease-disease associations when training the model.

5.2 Disease Clustering Methods

Many clustering algorithms are applied to disease-related data sets to gain insights into disease associations. These methods commonly utilize feature-vector representations of diseases and disease-related data [3, 5, 35]. Some clustering algorithms used for disease clustering include hierarchical, centroid-based, distribution-based, density-based, and deep learning-based clustering [3, 5, 35].

Hierarchical Clustering. Hierarchical clustering is a fundamental technique in bioinformatics for organizing complex biological data into meaningful structures. It represents data into tree-like structures (commonly a dendrogram) with either an agglomerative or divisive approach. For the agglomerative algorithm, each node is initially considered a separate cluster and then recursively merges two clusters with the least distance between them. Conversely, the divisive method initially assigns all data points into one cluster and then divides the sub-clusters until completion. These algorithms are simple to interpret due to the tree-based representation and are practical for smaller datasets, but they are sensitive to noise and outliers.

Centroid-Based Clustering. Centroid-based clustering algorithms categorize data based on its distance to the centroid of each cluster. K-means is a widely used algorithm among the centroid-based approaches. This algorithm partitions the data set into a predefined number of clusters k by assigning the data points to the nearest randomly initialized centroid. This process re-estimates the centroids for each iteration by calculating the mean of the points assigned to each cluster. Many prefer K-means due to its simplicity and efficiency, but it has a few limitations. Namely, the model assumes spherical and equally distributed clusters, which makes it less practical for complex cluster structures and sensitive to outliers.

Distribution-Based Clustering. Distribution-based clustering algorithms assume the underlying data distribution follows a particular statistical distribution model. These models can include Gaussian Mixture Models (GMM) and Expectation-Maximization (EM). These algorithms have advantages when modeling complex clusters, but they may have issues with overfitting, computational efficiency, and high-dimensional data.

Density-Based Clustering. Density-based algorithms identify clusters by analyzing areas with a high density of data points separated by sparse regions. DBSCAN and other density-based algorithms are insensitive to outliers and can identify clusters of any shape, but they have some disadvantages. Specifically, these algorithms fail to work with datasets of varying densities, and the computational cost can be significant with large datasets.

Deep Learning-Based Clustering. Deep learning-based clustering represents an advanced approach that applies deep learning techniques to the clustering task. These approaches often utilize neural network architectures, such as autoencoders (AE), variational autoencoders (VAE), and convolutional neural networks (CNN), to learn representations of high-dimensional or unstructured data [35]. For example, the deep convolutional autoencoder (CAE) presented by Sherif and Ahmed (2022) provided insights into six genetically similar virus strains of SARS-CoV-2, thus promoting the advancement of disease diagnosis and prevention [11]. These models are also capable of learning complex patterns and non-linear relationships. However, these models have a few disadvantages. Namely, these models can be computationally expensive and prone to overfitting.

Cluster Validation. The validation of cluster quality is a crucial step in disease clustering analysis to ensure that clusters are accurate and meaningful [5]. Common techniques that determine the cluster quality include external and internal validation. External validation involves comparing the clustering results to a standard or predefined knowledge. Alternatively, internal validation techniques assess cluster quality based on the data without relying on external information. These techniques evaluate intra-cluster and inter-cluster distances through methods like shortest path length and pairwise metrics. The shortest path length is measured by counting the edges between disease terms in hierarchical ontologies such as DO and MeSH [12, 16]. Cluster groups with the shortest mean path length indicate a higher degree of disease similarity, which suggests that disease pairs within the same cluster should have a smaller mean value than those in separate clusters. When using pairwise metrics, the mean value of the intra-cluster group should have a higher similarity score than that of the inter-cluster group. The study by Brady and Al-Mubaid (2024) illustrates internal and external cluster validation by utilizing shortest path length and disease-disease similarity scores from an external source to determine intra-cluster and inter-cluster qualities [10].

6   Conclusion

This paper reviewed the various aspects of disease similarity and disease clustering. We evaluated the resourcefulness of disease databases as information sources and their applications for semantic similarity measures and validation techniques. Furthermore, we acknowledged the limitations they present for this kind of research. This paper also discussed the disease attributes and computational methods used in disease similarity and clustering research. With this, many methodologies using novel disease-related data types, cutting-edge computational techniques, and integrated approaches have been and continue to be developed to quantify the associations between human diseases.

As this field progresses, several challenges and future trends become evident. There remains a critical need for standardization across disease resources to enhance data interoperability in disease similarity and clustering studies. Additionally, the integration of multiple data types presents both opportunities and challenges that require the advancement of current computational and validation methods. For future research, we recommend integrating deep learning techniques into disease similarity and clustering computational methods. Deep learning could be pivotal in mitigating the challenges associated with incomplete and inconsistent disease knowledge from databases. Investigating these challenges and future directions will further our ability to identify and understand the complex relationships between diseases.

Acknowledgements. This material is based upon work supported by the National Science Foundation under Grant No. 1928622.

Disclosure of Interests. The authors declare no competing interests that are relevant to the content of this paper.

References

1. Hodge, G.: Systems of knowledge organization for digital libraries: beyond traditional authority files. Digital Library Federation (2000)
2. Haendel, M.A., McMurry, J.A., Relevo, R., Mungall, C.J., Robinson, P.N., Chute, C.G.: A census of disease ontologies. Annu. Rev. Biomed. Data Sci. **1**, 305–331 (2018). https://doi.org/10.1146/annurev-biodatasci-080917-013459
3. Xiang, J., Zhang, J., Zhao, Y., Wu, F.-X., Li, M.: Biomedical data, computational methods and tools for evaluating disease–disease associations. Brief. Bioinf. **23**, bbac006 (2022). https://doi.org/10.1093/bib/bbac006
4. Cheng, L., et al.: Computational methods for identifying similar diseases. Molec. Therapy - Nucleic Acids **18**, 590–604 (2019). https://doi.org/10.1016/j.omtn.2019.09.019
5. Loftus, T.J., et al.: Phenotype clustering in health care: a narrative review for clinicians. Front. Artif. Intell. **5**, 842306 (2022). https://doi.org/10.3389/frai.2022.842306
6. Yang, X., et al.: Exploring novel disease-disease associations based on multi-view fusion network. Comput. Struct. Biotechnol. J. **21**, 1807–1819 (2023). https://doi.org/10.1016/j.csbj.2023.02.038
7. Su, S., Zhang, L., Liu, J.: An effective method to measure disease similarity using gene and phenotype associations. Front. Genet. **10**, 466 (2019). https://doi.org/10.3389/fgene.2019.00466
8. Rosário-Ferreira, N., Guimarães, V., Costa, V.S., Moreira, I.S.: SicknessMiner: a deep-learning-driven text-mining tool to abridge disease-disease associations. BMC Bioinf. **22**, 482 (2021). https://doi.org/10.1186/s12859-021-04397-w
9. Li, Y., Keqi, W., Wang, G.: Evaluating disease similarity based on gene network reconstruction and representation. Bioinformatics **37**, 3579–3587 (2021). https://doi.org/10.1093/bioinformatics/btab252
10. Brady, D., Al-Mubaid, H.: Disease clustering with process annotations from gene ontology (2024)
11. Sherif, F.F., Ahmed, K.S.: Unsupervised clustering of SARS-CoV-2 using deep convolutional autoencoder. J. Eng. Appl. Sci. **69**, 72 (2022). https://doi.org/10.1186/s44147-022-00125-0
12. Schriml, L.M., et al.: The human disease ontology 2022 UPDATE. Nucleic Acids Res. **50**, D1255–D1261 (2021). https://doi.org/10.1093/nar/gkab1063
13. Ashburner, M., et al.: Gene ontology: tool for the unification of biology. Nat. Genet. **25**, 25–29 (2000). https://doi.org/10.1038/75556
14. Consortium, T.G.O., et al.: The gene ontology knowledgebase in 2023. https://academic.oup.com/genetics/article/224/1/iyad031/7068118

15. Huntley, R.P., et al.: The GOA database: gene ontology annotation updates for 2015. Nucleic Acids Res. **43**, D1057–D1063 (2015). https://doi.org/10.1093/nar/gku1113

16. Leydesdorff, L., Comins, J.A., Sorensen, A.A., Bornmann, L., Hellsten, I.: Cited references and Medical Subject Headings (MeSH) as two different knowledge representations: clustering and mappings at the paper level. Scientometrics **109**, 2077–2091 (2016). https://doi.org/10.1007/s11192-016-2119-7

17. Amberger, J.S., Bocchini, C.A., Scott, A.F., Hamosh, A.: OMIM.org: leveraging knowledge across phenotype–gene relationships. Nucleic Acids Res. **47**, D1038–D1043 (2019). https://doi.org/10.1093/nar/gky1151

18. Köhler, S., et al.: The human phenotype ontology in 2021. Nucleic Acids Res. **49**, D1207–D1217 (2021). https://doi.org/10.1093/nar/gkaa1043

19. Zhu, Q., Nguyen, D.-T., Alyea, G., Hanson, K., Sid, E., Pariser, A.: Phenotypically similar rare disease identification from an integrative knowledge graph for data harmonization: preliminary study. JMIR Med. Inform. **8**, e18395 (2020). https://doi.org/10.2196/18395

20. Henry, V., et al.: INSIGHT-preAD study group: converting disease maps into heavyweight ontologies: general methodology and application to Alzheimer's disease. Database (Oxford) **2021**, baab004 (2021). https://doi.org/10.1093/database/baab004

21. Zhang, W., Zhang, H., Yang, H., Li, M., Xie, Z., Li, W.: Computational resources associating diseases with genotypes, phenotypes and exposures. Brief. Bioinform. **20**, 2098–2115 (2019). https://doi.org/10.1093/bib/bby071

22. Mathur, S., Dinakarpandian, D.: Finding disease similarity based on implicit semantic similarity. J. Biomed. Inf. **45**, 363–371 (2012). https://doi.org/10.1016/j.jbi.2011.11.017

23. Kumar, V., Chhabra, J.K., Kumar, D.: Impact of distance measures on the performance of clustering algorithms. In: Mohapatra, D.P., Patnaik, S. (eds.) Intelligent Computing, Networking, and Informatics: Proceedings of the International Conference on Advanced Computing, Networking, and Informatics, India, June 2013, pp. 183–190. Springer India, New Delhi (2014). https://doi.org/10.1007/978-81-322-1665-0_17

24. Bass, J.I.F., Diallo, A., Nelson, J., Soto, J.M., Myers, C.L., Walhout, A.J.M.: Using networks to measure similarity between genes: association index selection. Nat. Methods **10**, 1169–1176 (2013). https://doi.org/10.1038/nmeth.2728

25. Pedersen, T., Pakhomov, S.V.S., Patwardhan, S., Chute, C.G.: Measures of semantic similarity and relatedness in the biomedical domain. J. Biomed. Inf. **40**, 288–299 (2007). https://doi.org/10.1016/j.jbi.2006.06.004

26. Pesquita, C., Faria, D., Falcão, A.O., Lord, P., Couto, F.M.: Semantic similarity in biomedical ontologies. PLoS Comput. Biol. **5**, e1000443 (2009). https://doi.org/10.1371/journal.pcbi.1000443

27. Resnik, P.: Using information content to evaluate semantic similarity in a taxonomy (1995). https://arxiv.org/abs/cmp-lg/9511007v1

28. Lin, D.: An information-theoretic definition of similarity. In: Proceedings of the Fifteenth International Conference on Machine Learning, pp. 296–304. Morgan Kaufmann Publishers Inc., San Francisco (1998)

29. Jiang, J.J., Conrath, D.W.: Semantic similarity based on corpus statistics and lexical taxonomy. In: Chen, K.-J., Huang, C.-R., and Sproat, R. (eds.) Proceedings of the 10th Research on Computational Linguistics International Conference. pp. 19–33. The Association for Computational Linguistics and Chinese Language Processing (ACLCLP), Taipei (1997)

30. Zhu, F., et al.: Biomedical text mining and its applications in cancer research. J. Biomed. Inf. **46**, 200–211 (2013). https://doi.org/10.1016/j.jbi.2012.10.007

31. Zhang, N., Zang, T.: A multi-network integration approach for measuring disease similarity based on ncRNA regulation and heterogeneous information. BMC Bioinf. **23**, 89 (2022). https://doi.org/10.1186/s12859-022-04613-1

32. Koutrouli, M., Karatzas, E., Paez-Espino, D., Pavlopoulos, G.A.: A guide to conquer the biological network era using graph theory. Front. Bioeng. Biotechnol. **8**, 34 (2020). https://doi.org/10.3389/fbioe.2020.00034

33. Zhang, X.-M., Liang, L., Liu, L., Tang, M.-J.: Graph neural networks and their current applications in bioinformatics. Front. Genet. **12**, 690049 (2021). https://doi.org/10.3389/fgene.2021.690049

34. Gao, J., et al.: MTGNN: multi-task graph neural network based few-shot learning for disease similarity measurement. Methods **198**, 88–95 (2022). https://doi.org/10.1016/j.ymeth.2021.10.005

35. Karim, M.R., et al.: Deep learning-based clustering approaches for bioinformatics. Brief. Bioinf. **22**, 393–415 (2021). https://doi.org/10.1093/bib/bbz170

Composition Analysis and Identification
of Ancient Glass Products

Xuemei Yang$^{(\boxtimes)}$, Yuanyuan Zheng, Yanyan Xue, and Jianwei Xiao

School of Mathematics and Statistics, Xianyang Normal University, Xianyang 712000, China
yangxuemei691226@163.com

Abstract. Glass is a valuable proof of the initial transaction. Ancient glass buried underground is easily weathered by the environment, leading to changes in its composition ratio and affecting its classification. This article studies the composition analysis and identification of ancient Chinese glass products. Firstly, multiple correspondence analysis was used to study the correlation of surface (weathered or unweathered), decoration, type, and color. It was found that whether the glass surface is weathered is not related to type, but decoration and color, type and color are all related. Through non parametric testing, it was found that there are significant differences in the mean values of components such as silicon dioxide, potassium oxide, barium oxide, lead oxide, and calcium oxide among the four different categories of glass. Secondly, principal component analysis was performed separately on the composition for potassium and lead barium glass, and classify them into subcategories based on their main components, calcium oxide and lead oxide, and the rationality of the results was verified through cluster analysis. Then, Fisher discrimination, Bayesian discrimination, and principal component based Mahalanobis distance discrimination were used to predict the type of unknown glass samples, and the majority voting method was used to determine the type of unknown samples. Finally, for different categories of glass samples, Pearson correlation coefficient was used to analyze the correlation between chemical components. By comparison, it was found that there were significant differences in the correlation coefficients of certain components before and after weathering, indicating that surface weathering caused changes in the composition, which in turn led to changes in the correlation between components.

Keywords: multiple correspondence analysis · correlation analysis · principal component analysis · Potash glass · Lead barium glass

1 Introduction

The main components of glass are silica and other oxides. The birth of glass has a history of more than 4000 years, and the earliest discussion about ancient Chinese glass was in the 1930s. Historical data analysis suggests that ancient Chinese glass was introduced to China along the Silk Road in West Asia and Egypt, and then borrowed its technology and used local materials to produce glass that was similar but had different chemical compositions [1].

© The Author(s), under exclusive license to Springer Nature Switzerland AG 2024
H. Han (Ed.): SDSC 2024, CCIS 2158, pp. 78–94, 2024.
https://doi.org/10.1007/978-3-031-67871-4_6

Limestone serves as a stabilizer and is converted into calcium oxide (CaO) after calcination. Lead barium glass contains a significant amount of lead oxide (PbO) and barium oxide (BaO) due to the use of lead ore as a flux. Potassium glass, on the other hand, uses substances with potassium content as fluxes, resulting in a higher concentration of potassium oxide (K_2O). Adding different fluxes to different glasses results in different chemical compositions and concentrations [2].

Ancient glass is highly susceptible to weathering due to the influence of burial environments. After weathering, the color and chemical composition of glass can change, thereby affecting the correct judgment of its category. Is glass weathering related to the type of glass? What are the characteristics of the composition of potassium glass and lead barium glass respectively? How to classify subtypes based on these characteristics? What is the correlation between the composition of different types of glass? How does this correlation change after glass weathering? What changes occur in the composition of glass after weathering? How to predict the composition of glass before weathering based on its composition after weathering? How should we determine whether an unknown sample is potassium glass or lead barium glass?

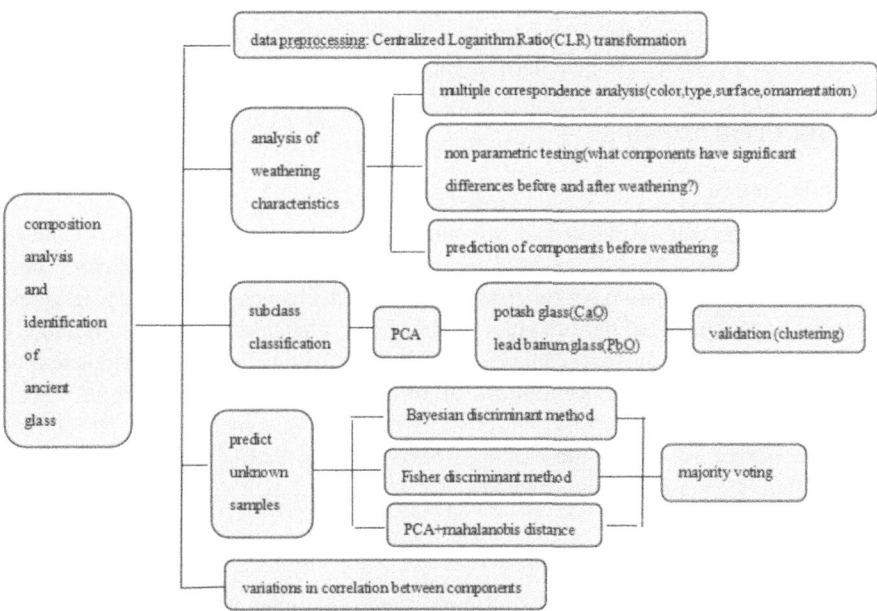

Fig. 1. Work flowchart. In our work, we employ multiple correspondence analysis to unveil the correlations among glass type, color, texture, and surface. PCA identifies the primary components and facilitates subclassification. Methods such as Bayesian and Fisher discriminant analyses are utilized to categorize unknown samples. Majority voting is adopted for decision-making. Ultimately, Pearson correlation analysis examines the correlations and differences among the components of various types of glass.

In this study, we use the method of multivariate statistical analysis to study the composition analysis and identification of ancient glass. Multiple correspondence analysis was used to explore the correlation between the types, colors, patterns, and surface weathering of glass. The non parametric testing method is used to reveal the changes in the composition of glass before and after weathering. The principal component analysis method was used to identify the main components of different types of glass and to classify them into subcategories. The correlation between the components of different types of glass is displayed through the Pearson correlation coefficient. We predict the composition of glass before weathering based on its ratio before and after weathering. To identify unknown glass samples, we used Bayesian discriminant analysis, Fisher discriminant analysis, specifically, we propose a principal component based Mahalanobis distance discriminant analysis, and use the majority voting method to obtain the final discriminant result. Our workflow diagram is as Fig. 1.

The rest of this paper is structured as follows. Section 2 introduces the data sources and preprocessing. Section 3 describes the correlation between the surface weathering of glass artifacts and their types, patterns, and colors, and predicts the composition concentration before weathering based on the composition characteristics of different types of glass. In Sect. 4, we use the main components to classify subcategories. Section 5 identifies the category of unknown samples. The last section compares the differences in the correlation between the chemical composition of different types of glass.

2   Data Preprocessing

2.1   Data Source

The experimental data in this article is from question C of the 2022 National College Student Mathematical Modeling Competition [3]. An excel file contains 3 sheets. Sheet 1 presents basic information for 58 glass samples, each characterized by four features: type(potash glass and lead barium glass), color(green, blue, etc.), texture(A, B and C), and surface(weathered and unweathered). Sheet 2 details the chemical composition of these samples, listing the concentrations of 14 chemical constituents, such as silicon dioxide and potassium oxide for each sample.The sampling points are random samples taken from the surface of each glass sample; Sheet 3 shows the 14 chemical composition proportion of 8 unclassified glass artifacts.

2.2   Data Preprocessing

We identified missing values across all three sheets. Samples with missing values in sheet 1 were directly removed, while those in sheets 2 and 3 were replaced with 0s. In sheet 2, some samples were labeled as weathered yet their sampling points were labeled as 'unweathered', treating these as unweathered. For two sampling points from the same sample, they were treated as distinct samples. Additionally, the data in sheets 2 and 3 represent component data, implying that the sum of each sample's component concentration should total 100%. However, measurement or other factors might cause deviations from this. Samples with component concentration sums between 85% and 105% were considered valid, while those (samples 15 and 17) outside this range were excluded.

2.3 Centralized Logarithm Ratio (CLR) Transformation

Due to the fact that the data is component data (the cumulative sum of the proportions of each component is 100%), such a fixed sum constraint can result in a negative bias of the covariance during the application of principal component analysis, thereby complicating interpretation, it is necessary to perform centralized logarithm ratio (clr) transformation on the data.

Let $X = (x_1, x_2, \cdots, x_p)$ be a p-dimensional vector, the centralized logarithm ratio (clr) transformation of X is

$$y_i = \ln \frac{x_i}{\sqrt[p]{\prod_{i=1}^{p} x_i}}, i = 1, 2, \ldots, p \tag{1}$$

The inverse transformation is

$$x_i = \ln \frac{\exp(y_i)}{1 + \sum_{i=1}^{p-1} \exp(y_i)}, i = 1, 2, \ldots, p \tag{2}$$

Centralized Logarithm Ratio transformation can make the transformed components correspond one-to-one with the original components, enhancing interpretability [4–6].

3 Correlation Analysis of Weathering Characteristics and Component Prediction

3.1 Multiple Correspondence Analysis Between Surface, Decoration, Type and Color

Since the four variables of glass type, color, surface, and decoration are qualitative variables, we use the method of Multiple Correspondence Analysis(MCA) to analyze their correlation. Correspondence analysis is a multivariate statistical analysis method developed on the basis of R-type and Q-type factor analysis, also known as R-Q type factor analysis. It uses appropriate scaling methods for the original data, and simultaneously performs R-type and Q-type factor analysis. Two common factors are taken as the coordinate axes of the two-dimensional plane, and variables and samples are classified together on the same factor plane to reveal the internal connections between the studied samples and variables [7, 8]. The results of multiple correspondence analysis are shown in Figs. 2.

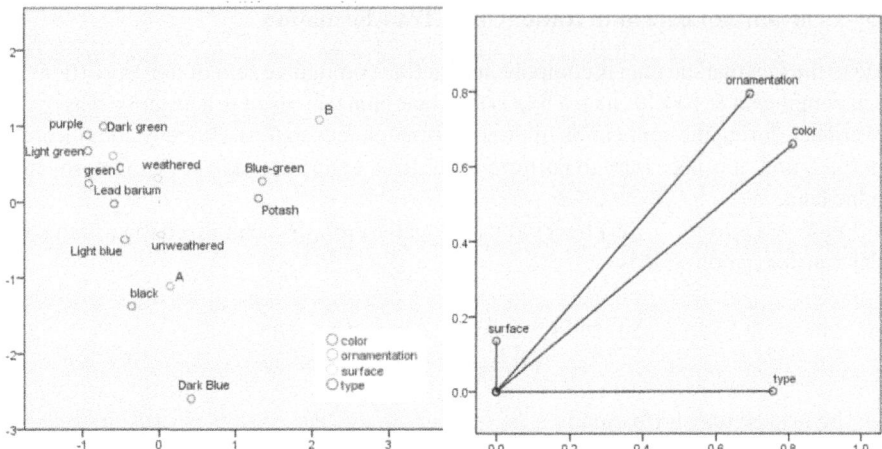

Fig. 2. Results of multiple correspondence analysis. The left graph is a joint graph of category points. The right graph is discrimination metric. In the left figure, each point represents the values of four variables, and the distance between points represents their level of intimacy. The angle between the line segments in the right figure reflects the correlation between variables. (Color figure online)

The values of each variable are reflected as points in the joint graph of category points, the distance between points represents their closeness. That is, the distribution of closely related category points is relatively concentrated, while the distribution of loosely related category points is relatively scattered. It can be seen that the glass sample of C decoration are mostly green, light green, dark green, and purple; The distance between lead barium glass and weathered, as well as the distance between lead barium glass and unweathered are the same, the distance between potassium glass and weathered, as well as the distance between potassium glass and unweathered are also the same, indicating that the type of glass is not related to whether the surface is weathered; Lead barium glass is mostly light blue and green, while potassium glass is mostly blue-green; There are very few glass with dark blue and B-type decorations.

The discriminant metric graph uses the angle between line segments to represent the degree of closeness of variables. The smaller the angle, the closer the correlation between them. As shown in the right graph of Fig. 2, the closest correlation is between decoration and color, followed by decoration and surface, and then color and type, while the correlation between surface and type is not significant.

3.2 Statistical Rules of Chemical Composition Concentration in Glass Samples

In order to study the changes in composition of different types of glass before and after weathering, we divided the glass samples into four categories: potassium weathering, lead barium weathering, potassium unweathering, and lead barium unweathering. K-S test was used to test the distribution shape of components in each category, and it was found that most of the components were not normally distributed. Therefore, non parametric test was used to compare the differences in the mean values of the components in various samples [9]. We have selected the following list of significant differences in the mean composition before and after weathering at a significance level of 0.05.

Table 1. Non parametric test results

	potassium glass					lead barium glass				
	K_2O	CaO	MgO	Al_2O_3	Fe_2O_3	SiO_2	Na_2O	CaO	PbO	P_2O_5
ManWhitney	6	16	9	1	12	43	192	149	62	116
Wilcoxon	34	44	37	29	40	368	517	425	338	392
Z score	−3.05	−2.20	−2.90	−3.47	−2.55	−5.05	−2.47	−2.86	−4.65	−3.56
Significance	0.00	0.03	0.00	0.00	0.01	0.00	0.01	0.00	0.00	0.00

From Table 1, it can be seen that there are significant differences in the five components of potassium oxide, calcium oxide, magnesium oxide, aluminum oxide, and iron oxide in potassium glass before and after weathering. There are significant differences in silicon oxide, sodium oxide, calcium oxide, lead oxide, and phosphorus oxide in lead barium glass before and after weathering. We further plot the mean values of each component of the various samples as shown in Fig. 3.

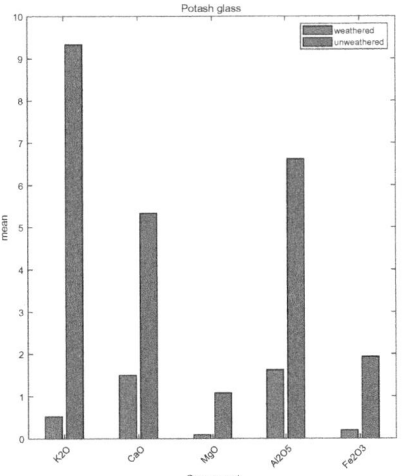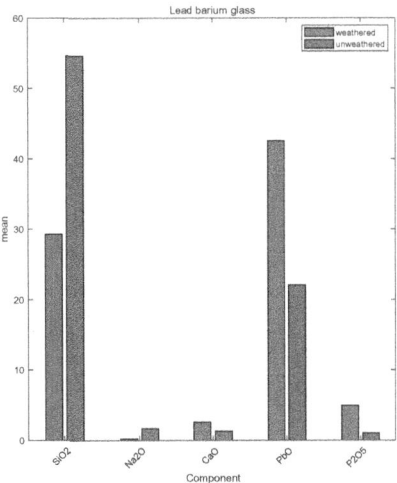

Fig. 3. Comparison of Mean Values of Main Components. The left figure shows that the concentration of five components, including potassium oxide and calcium oxide, in potassium glass significantly decreases after weathering; The right figure shows that the concentration of silicon oxide and sodium oxide in lead barium glass also decreases after weathering, but the concentration of calcium oxide, lead oxide, etc. increases after weathering.

It can be seen that the concentration of potassium oxide, calcium oxide, magnesium oxide, aluminum oxide, and iron oxide in potassium glass is greatly reduced after weathering. In fact, the silicon dioxide concentration in weathered potassium glass increases, ensuring that the total proportion of all component remains unchanged. The concentration of silicon oxide and sodium oxide in lead barium glass decreases after weathering, while the concentration of calcium oxide, lead oxide, and phosphorus oxide increases after weathering.

3.3 Predicting the Chemical Composition of Glass Samples Before Weathering

From the above analysis, it can be seen that the composition content of glass will change after weathering, so it is necessary to predict the content of a certain component of glass before weathering. We assume that the ratio of the concentration of a certain chemical component in glass before and after weathering is constant, then a model is established.

$$c_1 = c_2 \times \frac{\mu_1}{\mu_2} \tag{3}$$

Here c_1 and c_2 represent the concentration of a certain chemical component in glass before and after weathering, respectively, while μ_1 and μ_2 represent the mean concentration of this component before and after weathering respectively. The predicted component concentration in glass sample before weathering is as follows (Table 2).

Table 2. Prediction results of component concentration before weathering

Sampling points	SiO2	K2O	CaO	Al2O3	PbO	BaO	P2O5
02	108.4595	159.0313	1.9122	7.3615	0	0	0.3154
07	144.9099	0	33.2274	3.7088	0	0	0.1644
08	136.1806	82.6199	3.2196	7.6050	0.1041	0	0.1779
08–1	96.5388	196.9186	2.6639	10.3023	0.5871	1.8249	0.1887
09	148.6488	9.3922	14.3336	2.4726	0	0	0.0943

All predicted results can be found in the supplementary materials (https://pan.baidu.com/s/1e6cRUesRHE-OuZW1mHOt6w, extract code: qwbf).

4 Subclass Classification

In order to classify potassium glass and lead barium glass into subclasses, we need to know their respective principal components.

4.1 Principal Component Analysis (PCA) of Lead Barium Glass and Potassium Glass

PCA is a statistical method aimed at transforming a set of potentially correlated variables into a set of linearly uncorrelated variables (principal components) through orthogonal transformation. The main goal of PCA is to find the components that can explain the most variance in the data, which are linear combinations of the original features. The core idea of PCA is to use the idea of dimensionality reduction to transform multiple indicators into a few comprehensive indicators, in order to reduce the number of variables while preserving the information of the original dataset as much as possible [10].

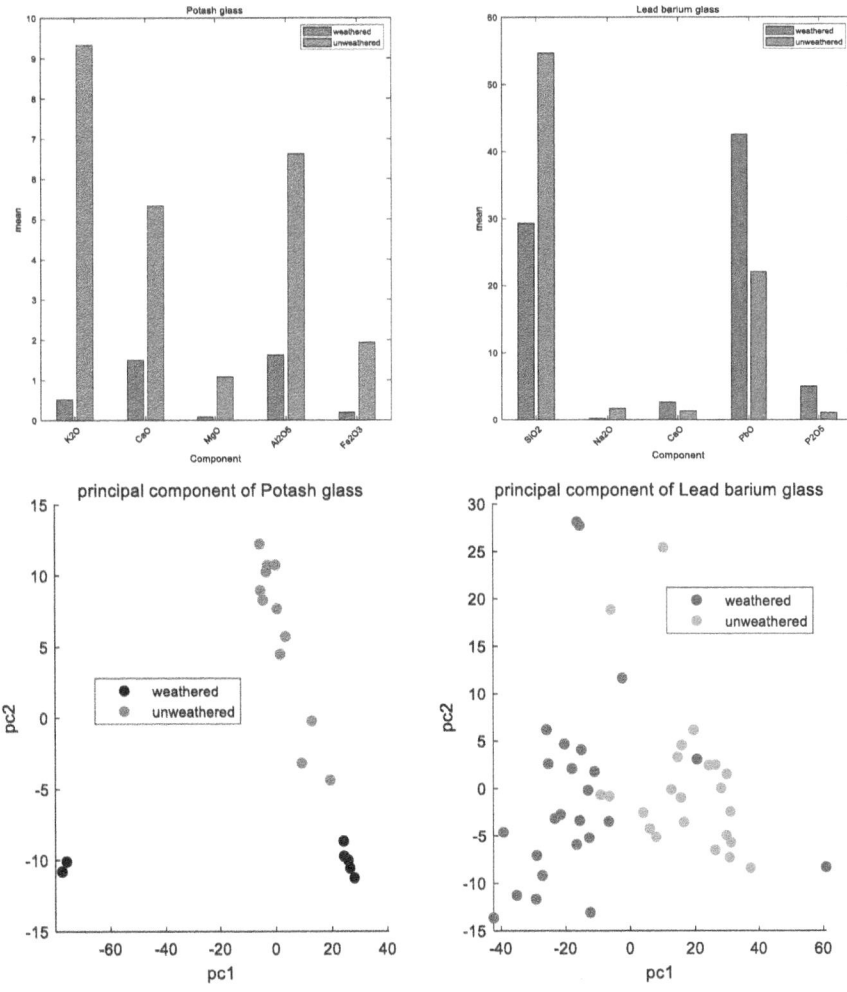

Fig. 4. Principal components of the glass sample. It can be seen that whether it is potassium glass or lead barium glass, their samples before and after weathering are clearly divided into two clusters.

PCA was performed on the component of potassium glass and lead barium glass respectively, for each sample, we take the first two principal component scores to plot the points, and the results are as follows.

As shown in Fig. 4, the second principal component score of the unweathered sample in potassium glass is higher, while the second principal component score of the weathered sample is lower. This is because potassium oxide is the second principal component of potassium glass (Table 3), and as shown in Fig. 4, the potassium oxide in potassium glass greatly decreases after weathering. The first principal component score of the unweathered sample in lead barium glass is higher, while the first principal component score of the weathered sample is relatively lower, This is because silicon oxide is the first principal component of lead barium glass (Table 4), and as shown in Fig. 4, the silicon oxide in lead barium glass is greatly reduced after weathering, and there is no difference in their second principal component scores.

It can be clearly observed that there are two outliers in the potassium weathering samples, numbered 08 and 26, which are labeled as 'severe weathering points'. Observing their main chemical composition, it is found that the concentration of silicon dioxide is much lower than that of other samples in the same category, while the concentration of lead oxide, barium oxide, and sulfur oxide is much higher than that of other samples in the same category, so they are separated from other samples.

Table 3. Coefficients of Principal Components of Potassium Glass

chemical composition	The first three principal component coefficients of potassium glass			characteristic value	Accumulated proportion
SiO2	**0.88178690**	−0.33632886	−0.02999	876.6542072	89.70%
Na2O	−0.0016397	0.06239794	0.062026	83.45045726	98.23%
K2O	0.00541932	**0.54966399**	−0.4650	7.931430886	99.05%
CaO	−0.0262389	0.27743568	**0.71198**	4.282766181	99.48%
MgO	0.00195007	0.04755185	−0.1013	2.379645843	99.73%
Al2O3	0.00597976	0.28286858	−0.17608	1.042047594	99.83%
Fe2O3	−0.0000297	0.10572883	0.17171	0.703460957	99.91%
CuO	−0.0208064	0.03753364	0.35036	0.448089562	99.95%
PbO	−0.3026477	−0.21333456	−0.05540	0.281222241	99.98%
BaO	−0.3209462	−0.23926568	−0.07131	0.127916747	99.99%
P2O5	−0.0630540	−0.03785320	−0.20809	0.042598769	100.00%
SrO	−0.0056036	−0.00574309	−0.01206	0.013074811	100.00%
SnO2	0.00190034	−0.00036197	−0.16234	0.004526574	100.00%
SO2	−0.1504855	−0.20918737	−0.02498	0.000148958	100.00%

Table 4. Coefficient of Principal Components of Lead Barium Glass

chemical composition	The first three principal component coefficients of Lead Barium glass			characteristic value	Accumulated proportion
SiO2	**0.77579659**	−0.39917419	−0.31559210	568.123064	81.36%
Na2O	0.02427460	0.01488906	−0.05117136	87.7587119	93.93%
K2O	0.00092939	0.00011600	0.00699798	25.8012170	97.62%
CaO	−0.03108505	−0.01730461	0.18746721	7.83116215	98.74%
MgO	0.00034619	−0.01247724	0.07399853	3.27212046	99.21%
Al2O3	0.04705517	−0.005204	0.32722506	2.00116386	99.50%
Fe2O3	0.00046215	−0.01142577	0.07225203	1.92174391	99.78%
CuO	−0.01875417	0.20727115	−0.07032995	0.59545683	99.86%
PbO	**−0.61990741**	−0.46934754	−0.40749237	0.3368870	99.91%
BaO	−0.05681830	**0.68663180**	−0.49882876	0.29949798	99.95%
P2O5	−0.08043448	0.0127651	**0.56960172**	0.20931058	99.98%
SrO	−0.00573699	0.00045545	0.00086403	0.05956243	99.99%
SnO2	0.00070618	0.00074088	0.00861387	0.03881970	100.00%
SO2	0.00059521	0.0328725	-0.02168533	0.03054086	100.00%

Table 3 shows the coefficients of the first three principal components of potassium glass, with the bolded numbers being the larger coefficients, indicating that the first principal component of potassium glass represents silicon dioxide, the second principal component represents potassium oxide, and the third principal component is calcium oxide.

According to the cumulative proportion of eigenvalues, selecting the first three principal components can result in a cumulative contribution rate of 99.05%, indicating that the three common factors can explain about 99% of the total variance. The chemical composition concentration of calcium oxide can serve as the standard for classifying potassium glass.

Table 4 shows the coefficients of the first three principal components of of lead barium glass, with the bolded numbers being the larger coefficients, indicating that the first principal component of lead barium glass represents silicon oxide and lead oxide, the second principal component represents barium oxide, and the third principal component is phosphorus oxide.

According to the cumulative proportion of eigenvalues, selecting the first three principal components can result in a cumulative contribution rate of 97.62%, indicating that the three common factors can explain about 98% of the total variance. The chemical composition concentration of lead oxide can serve as the standard for the classification of lead barium glasses.

4.2 Subclass Classification

Through the above principal component analysis, we know that calcium oxide is the main component of potassium glass and lead oxide is the main component of lead barium glass, they can be used as the standard for subclassing [11].

Different subclasses of glass have different formulas. The first type G1 and the second type G2 both belong to potassium glass, but the concentration of CaO (>6%) in the first type glass is significantly higher than that in the second type glass; The last three types of G3, G4, and G5 all belong to lead barium glass, but the concentration of PbO shows an increasing trend. The concentration of SiO2 in G3 is higher, while the concentration of PbO is lower (<25%); The concentration of PbO in G5 is relatively high (>40%), which is higher than the concentration of SiO2. According to the principles of subclassing, all samples can be divided into the 5 categories as Table 5.

Table 5. Criteria and Results for Subclassing

Type		Distinguishing feature	Composition	Results
potassium glass	G1	CaO > 6%	K2O-CaO-SiO2	01,04,05,13,14,16
	G2	CaO <= 6%	K2O-SiO2	03,07,09,10,12,18,21,22,27
lead barium glass	G3	PbO <= 25%	PbO-BaO-SiO2	20,23,28,29,31,32,33,35,37,42,44,45,48,49,53,55
	G4	25% < PbO < 40%	PbO-BaO-SiO2	08,11,24,25,26,30,46,47,49,58
	G5	PbO >= 40%	CaO-PbO-BaO-SiO2	02,19,34,36,38,39,40,41,43,50,51,52,54,56,57

To verify the rationality of the above classification, cluster analysis was conducted on the samples. Table 6 shows the clustering results of some samples.

The clustering results of other samples can be found in the supplementary materials, and the clustering results are consistent with the results of classification.

Table 6. Cluster Members

Case number	Sampling points	category	Distance to class center
1	7	3	4.109
2	9	3	5.797
3	10	3	7.601
4	12	3	4.995
5	22	3	3.256
6	27	3	3.706
7	1	4	19.957
8	03–1	3	4.503

5 Determine the Type of Unknown Samples

In order to determine the category of 8 given unknown samples, we use 56 known samples as training samples and use three methods for discrimination: Bayesian discrimination, Fisher discrimination, and principal component based Mahalanobis distance discrimination.

Table 7. Bayesian and Fisher discriminant results

Sampling points	P1	P2	Bayesian discrimination results	Fisher discriminant function value	Fisher discrimination result
A1	0.08283	0.91717	Lead barium glass	−0.73259	Potash glass
A2	0.00003	0.99997	Lead barium glass	0.96694	Lead barium glass
A3	0.00026	0.99974	Lead barium glass	0.4886	Lead barium glass
A4	0.00001	0.99999	Lead barium glass	1.15014	Lead barium glass
A5	0.00008	0.99992	Lead barium glass	0.72001	Lead barium glass
A6	0.99976	0.00024	Potash glass	−2.97119	Potash glass
A7	0.99442	0.00558	Potash glass	−2.31142	Potash glass
A8	0.02297	0.97703	Lead barium glass	−0.45257	Potash glass

5.1 Bayesian Discrimination and Fisher Discrimination

The principle of Bayesian discriminant analysis is Bayesian thinking, which assumes that the prior probability of the research object is known, then the prior probability distribution is corrected with the samples to obtain the posterior probability distribution, and classify the sample into the category with the highest posterior probability [9].

The Fisher discriminant method involves projecting all training samples(vector) into a certain direction, using the idea of one-way analysis of variance to separate the projected groups [12, 19], and then using distance discriminant method to distinguish them. Table 7 are the results of Bayesian discriminant method and Fisher discriminant method. P1 represents the posterior probability belonging to potassium glass, P2 represents the posterior probability belonging to lead barium glass.

5.2 Principal Component based Mahalanobis Distance Discrimination

If the mean of the i-th class training sample is μ_i and the covariance matrix is Σ_i, then the Mahalanobis distance from the sample to the i-th class is

$$d_i = (X - \mu_i)' \Sigma_i^{-1} (X - \mu_i) \tag{4}$$

The principal component based Mahalanobis distance discrimination method refers to performing principal component transformation on training and testing samples, calculating the first three principal component scores, replacing the original data with principal component scores, calculating the Mahalanobis distance from unknown samples to the two classes, and grouping the samples into the nearest class. The results are shown in the Table 8.

Table 8. Results of Mahalanobis distance discrimination based on principal components

Sampling points	A1	A2	A3	A4	A5	A6	A7	A8
type	Lead barium glass	Lead barium glass	Lead barium glass	Lead barium glass	Lead barium glass	Potash glass	Potash glass	Lead barium glass

Table 9. Discriminant Results of Majority Voting

Sampling points	Number of votes for potassium glass	Number of votes for lead barium glass	Type
A1	1	2	Lead barium glass
A2	0	3	Lead barium glass
A3	0	3	Lead barium glass
A4	0	3	Lead barium glass
A5	0	3	Lead barium glass
A6	3	0	Potash glass
A7	3	0	Potash glass
A8	1	2	Lead barium glass

5.3 Majority Voting

Due to the differences in the three results, the minority obeys the majority, and the majority voting method is used to make the final decision [14, 15], the result is shown in Table 9.

The above method was used to judgment the 56 known samples, with an accuracy rate of 95.52%.

6 The Correlation Between the Chemical Components of Various Samples and Its Differences

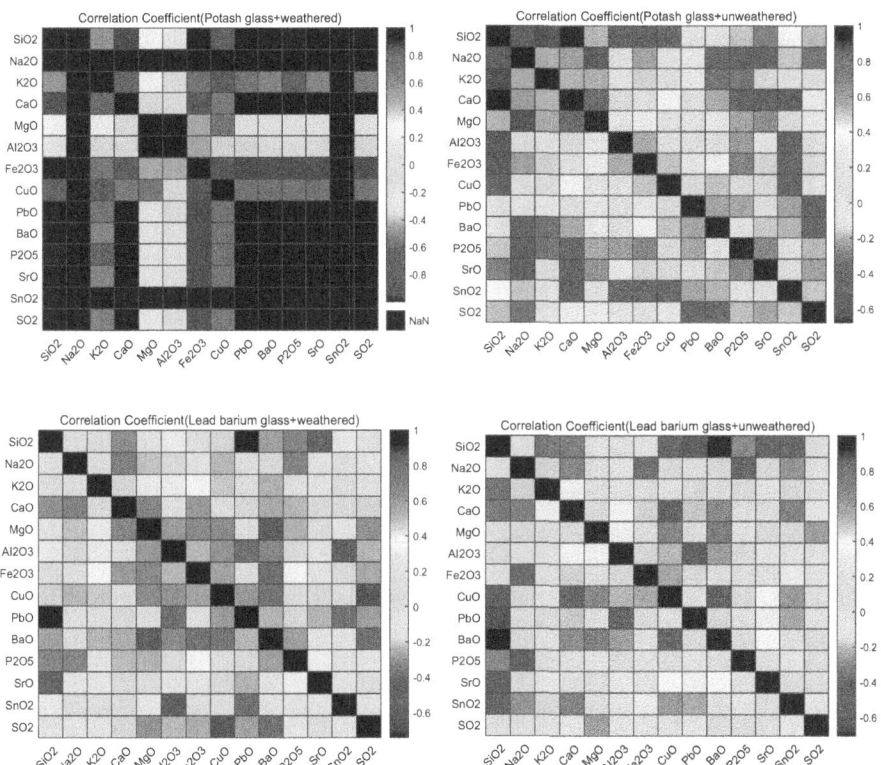

Fig. 5. Correlation coefficient heatmap. It can be seen that the two images in the first row are significantly different, and the two images in the second row are also significantly different. This indicates that the correlation between the components of both potassium glass and lead barium glass has changed after weathering.

6.1 Pearson Correlation Analysis

We have calculated the Pearson correlation coefficient [13] matrix of the four types of glass: potassium weathered, lead barium weathered, potassium unweathered, and lead barium unweathered, which is represented by a heatmap as Fig. 5. Due to the absence of Na2O and SnO2, potassium weathered glass exhibits black areas; Most of the absolute values of the correlation coefficients of potassium weathered glass are relatively large, such as the strong positive correlation between CaO and various components; There is also a strong positive correlation between PbO, BaO, P2O5, SrO, and SO2; SiO2 has a strong negative correlation with various components; Fe2O3 has a strong negative correlation with various components; And these correlations are slightly weaker before weathering; In lead barium weathered glass, there is a strong positive correlation between CaO, MgO, Al2O3, and Fe2O3, but these correlations are also slightly weaker before weathering.

6.2   Differences in Dorrelation

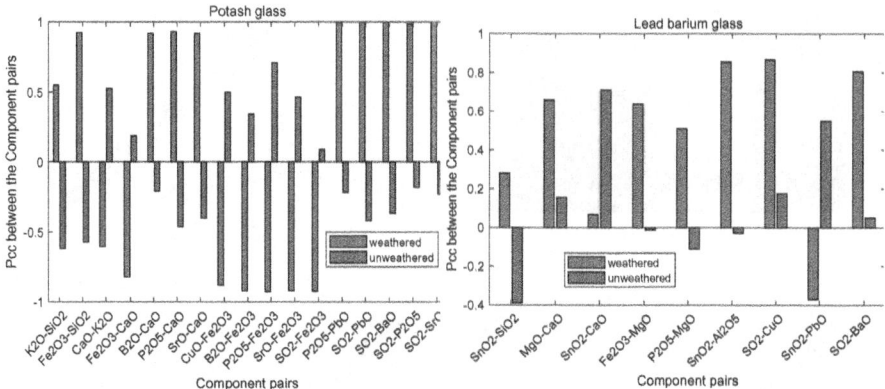

Fig. 6. Comparison of correlation. We selected 17 pairs of correlation coefficients with significant changes in potassium glass and 9 pairs of correlation coefficients with significant changes in lead barium glass. It can be seen that some of these correlation coefficients changed from positive to negative, while others were exactly the opposite.

Figure 6 shows that the correlation coefficient between the components of Potash glass varies greatly before and after weathering, with 17 pairs having a difference greater than 1, as shown in the left Figure; Among them, 10 pairs changed from negative correlation before weathering to positive correlation after weathering, and 7 pairs changed from positive correlation before weathering to negative correlation after weathering; The correlation coefficients between the components of lead barium glass vary relatively little before and after weathering, with 9 pairs having differences greater than 0.5, as shown in the Figure on the right. Among them, 4 pairs have changed from negative correlation before weathering to positive correlation after weathering, 1 pair has changed from positive correlation before weathering to negative correlation after weathering, 3 pairs have increased positive correlation coefficients after weathering, and 1 pair has decreased positive correlation coefficients after weathering, such as SnO2-CaO.

7   Conclusions and Future Work

We used multivariate statistical analysis to study the composition characteristics and category identification of ancient Chinese glass. Discovering that weathering is not related to type, but there is a correlation between pattern and color, type and color; In addition to SiO2, calcium oxide and lead oxide are the main components of potassium glass and lead barium glass, respectively, and can serve as the basis for subclassing; After weathering, the composition of glass will change, and the correlation between components will also change; Experiments have shown that Bayesian discrimination and principal component based Mahalanobis distance discrimination can effectively classify glass.

However, when predicting the content of glass components before weathering, we consider the ratio of a specific component of glass content before and after weathering as a constant value, and use the product of the content of glass components after weathering and this ratio to predict the content of glass components before weathering. While this approach is simple and feasible, it ignores the distribution state of glass component content itself and the changes in this distribution state after weathering [16, 17]. In order to more accurately predict the composition content of glass before weathering, we should also consider the differences in the correlation between different components before and after weathering, in addition, when searching for the main components of glass, we can also consider using factor analysis method [18, 20], which is the direction for our future improvement.

References

1. Chen S.: Archaeologist julian huntson lectures on cultural exchange on the silk road from ancient glassware. Popular Archaeol. **06**, 89 (2014)
2. Zhao, Z.: A Study on the Composition System and Manufacturing Technology of Glass Beads Excavated from the Shizigou Site Group in Balikun. Shanxi:Northwestern University, Xinjiang (2016)
3. http://www.mcm.edu.cn/html_cn/node/388239ded4b057d37b7b8e51e33fe903.html
4. Ao, B.: A Study on Anomaly Detection Method for Component Data Based on MRCD. Zhejiang University of Finance and Economics, Zhejiang (2023)
5. Wang, H.: Linear regression model for component data. System Eng. **21**(2), 102–106 (2003)
6. Zhang, C.: Principal component analysis of component data and its applications. Math. Stat. Manag. **15**(3), 11–14, 48 (1996)
7. Ruan, J., Ji, H.: Practical SAS Statistical Analysis Tutorial. China Statistical Publishing House, Beijing (2013)
8. Chen, T.: Quantitative Archaeology. Peking University Press, Beijing (2005)
9. Li, H., Li, Y., Zheng, X.: SPSS Statistical Experiment Tutorial. Tsinghua University Press, Beijing (2015)
10. Xue, X., Jia, Q., Chen, Z.: Comprehensive evaluation of agronomic traits of recombinant inbred lines of tartary buckwheat based on principal component analysis. Zhejiang Agric. J. **4**, 1–13 (2024)
11. Fu, X.: A study on the composition of ancient glass from southern and southwest china based on multivariate statistical analysis. Cult. Relics Prot. Archaeol. Sci. 18(4), 6–13 (2006)
12. Kong, W.: Research on the Recognition and Classification Method of Jujube Images in Southern Xinjiang Based on Hyperspectral Technology. Tarim University, Xinjiang (2017)
13. Luo, G., Tao, Y., Li, C.: Correlation analysis of the in vitro inhalation performance of Re Du Ning inhalation solution. Chin. J. Exp. Formulas **24**(04), 1–7 (2018)
14. Han, H., Teng, J., Xia, J., Wang, Y., Guo, Z., Li, D.: Predict high-frequency trading marker via manifold learning. Knowl.-Based Syst. **213**, 106662 (2021)
15. Han, H., Li, W., Wang, J., Qin, G., Qin, X.: Enhance explainability of manifold learning. Neurocomputing **500**, 877–895 (2022)

16. Han, H.: Derivative component analysis for mass spectral serum proteomic profiles. BMC Med. Genomics **7**, 1–14 (2014)
17. Han, H., Li, D., Liu, W., Zhang, H., Wang, J.: High dimensional mislabeled learning. Neurocomputing **573**, 127218 (2024)
18. Yang, X., Han, H.: Factors analysis of protein O-glycosylation site prediction. Comput. Biol. Chem. **71**, 258–263 (2017)
19. Shawe, T., Nello, I.: Kernel Methods for Pattern Analysis. Machinery Industry Press, Beijing (2006)
20. Anderson, T.W.: A Introduction to Multivariate Statistical Analysis, 2nd edn. World Publishing Co. Wiley, New York (1984)

Data Entropy-Based Imbalanced Learning

Yutao Fan[1,2,3,4](✉) and Heming Huang[1,2,3]

[1] Qinghai Normal University, Xining 810008, China
`fanyutao2005@126.com, huanghm@qhnu.edu.cn`
[2] State Key Laboratory of Tibetan Intelligent Information Processing and Application,
Xining 810008, China
[3] Key Laboratory of Tibetan Information Processing, Ministry of Education,
Xining 810008, China
[4] North China Institute of Science and Technology, Beijing 065201, China

Abstract. All the time the skewness of observations is thought as the reason of poor classification performance, especially the bias in classification performance among classes in machine learning. However, our recent study challenges this notion. We argue that the bias of classification performance comes from the imbalance of information of classes rather than just that of observations. To reflect the information imbalance of classes, we propose an indicator data entropy that captures the randomness within classes. A dataset with balanced and higher data entropies across its classes is more likely to exhibit improved classification performance. Furthermore, we propose another indicator data mutual information that quantifies the similarity between classes. Higher values indicates that the models can leverage learning from classes to enhance learning capacity. Therefore, reducing the difference in data entropy between classes and enhancing data mutual information concurrently is advantageous for classification. Our experiments, conducted across four models SVM, CNN, Transformer (including its variants ViT), and DNN, on datasets CIFAR-10, Airline Satisfaction, Smoking Body Signal and Liver Cirrhosis, validate the efficacy of our proposed indicators. Through rebalancing the data entropy distribution among classes and increasing the data entropy within classes as well as the data mutual information in the Liver Cirrhosis dataset using resampling techniques, we observe classification enhancements measured in d-index across four models.

Keywords: data entropy · deep learning · imbalanced learning

1 Introduction

In real-world applications, datasets often exhibit imbalanced distribution, where certain class levels are severely underrepresented. In such cases, learning models have shown bias towards the majority class, impeding accurate predictions for the minority class [1].

The imbalance ratio, which represents the ratio of observations between the majority and minority classes, is commonly used to characterize the skewed distribution of data [2]. This ratio can be further categorized into inter-class and intra-class imbalances.

H. Han (Ed.): SDSC 2024, CCIS 2158, pp. 95–109, 2024.
https://doi.org/10.1007/978-3-031-67871-4_7

Moreover, most researcher think that the imbalanced observations of different classes are the primary reason why the poor classification performance, especially low F1 scores of minority classes.

Consequently, to address the challenges posed by imbalanced datasets, various methods are proposed, including resampling techniques such as over-sampling [3] and under-sampling [4], data augmentation strategies like using GANs and its variants [5], and other innovative approaches. The primary objective of these methods is to reduce the class imbalance ratio towards 1. However, it is essential to recognize that the bias introduced by learning models is not solely determined by the distribution of observations.

In this study, we found that the skewness of information among different classes in a dataset is the deeper reason for bias in classification tasks. Thus, we propose two indicators, that is, data entropy and data mutual information to reflect the imbalance of information.

Data entropy is employed to assess the diversity and variance within the data. Higher data entropy values signify increased diversity and complexity, meaning more information embedded in datasets and posing greater challenges for learning algorithms. As a result, it is the imbalance of data entropy of classes that lead to the bias in classification performance.

Conversely, data mutual information quantifies the shared information between two datasets. Increased values of data mutual information indicate that learning models can effectively utilize features from both datasets, enhancing the overall performance.

We carry out two groups of experiments on four models and four datasets to validate the two indicators proposed. Experimental results show that the two indicators can reflect the skewness of data information of classes and thus bias in classification performance can be improved through balancing the data entropy of classes.

The rest of this paper is structured as follows. In Sect. 2 and Sect. 3, we formally define the concepts of data entropy and data mutual information, and as well as their key characteristics. Section 4 presents the introduction to machine learning models or classifiers used in our experiments and outlines their efficacy in handling imbalanced data distribution. Section 5 introduces the two performance evaluation metrics used in classification tasks. Section 6 details the experimental analysis, followed by a comprehensive discussion of the experimental results. Section 7 makes the conclusions from our study, outlining the limitations of our proposed approach and further enhancements in the future work.

2 Data Entropy

In machine learning, the goal of classification tasks is to create a model or a learning algorithm to predict what class the new observations belong to according to obtained features and labels from training dataset through training [6].

Given a dataset $\mathbb{X} = \{(x_1, y_1), (x_2, y_2), ..., (x_m, y_m)\}, x_i (i = 1, 2, ...m)$ represents an observation (each x_i has n features) and $y_j (i = 1, 2, ...m)$ represents the label associated with each observation.

The model f aims to predict the label $\hat{y}_{new}$ for a new observation x_{new} by what learned from the training datasets. Obviously, it is essential for the model f to learn diverse and informative features from the training sets in order to enhance the performance of the classification task.

Singular values from Singular Value Decomposition (SVD) are very helpful in analyzing the diversity and extracting information in datasets for its capabilities in capturing the underlying structure and patterns [7, 8]. Larger singular values indicate that the corresponding singular vectors contribute more significantly to the data variability, which means the dataset has more varieties. The distribution of singular values can provide insights into the information content of the dataset. If the singular values of a dataset consist of a few dominant singular values with large magnitudes and many small singular values, then the dataset have comparatively low information, leading to easy classification.

In information theory, Shannon entropy is a typical indicator used to quantify the amount of uncertainty or disorder in a dataset [9, 10]. For a dataset, high entropy values indicate a more diverse and balanced dataset and more randomness and information embedded in the dataset, while low entropy values suggest class imbalance that may require attention in classification tasks [11]. Researchers have proposed various approaches such as category entropy [12] and asymmetric entropy [13], based on Shannon entropy to help improving classification performance.

We employed a novel data entropy measure proposed by Han et *al.* for assessing the diversity and information of a dataset [8],

$$\mathcal{H}(\mathbb{X}) = -\sum_{i=1}^{n} u_i \log_2 u_i \qquad (1)$$

where $u_i = \frac{s_i}{\sum_{i=1}^{n} s_i}$, and s_i is the i th singular value of $\mathbb{X}$ and n is the number of singular values of $\mathbb{X}$. Especially, when the number of observations of $\mathbb{X}$ is more than that of features of $\mathbb{X}$, n is the number of features of $\mathbb{X}$.

From the Eq. 1, we observe that the magnitude of data entropy is primarily influenced by the number of singular values and is independent of the observations from the classes in the dataset. In fact, when carrying out classification tasks, the root reason for better classification comes from whether obtaining enough information from features to differentiate between classes.

On one hand, a higher data entropy, reflecting a wider distribution of singular values and increased variability in the underlying patterns of the dataset . This increased variability means more randomness and information which can be provided to the learning model to distinguish between the classes to learn the information fully, the classifiers should have stronger learning abilities.

On another hand, in the classification tasks, a significant difference between different parts of the dataset $\mathbb{X}$ show the presence of learning difference for learning models, leading to bias of classification performance. Consequently, decreasing difference between the data entropy of classes will benefit the overall classification performance.

3 Data Mutual Information

When carrying out classification on a dataset, the shared information between classes can help in extracting common discriminative features essential for distinguishing between classes. This shared information enhances the representation of the data, leading to more accurate classification results [14]. In line with this viewpoint, we proposed another metric called data mutual information, based on singular values, to assess the classification performance. The definition of data mutual information is shown in Eq. 2.

$$\mathcal{I}(\mathbb{A}; \mathbb{B}) = \mathcal{H}(\mathbb{A}) + \mathcal{H}(\mathbb{B}) - \sum_{i=1}^{n} u_i v_i \log u_i v_i \tag{2}$$

where $u_i = \frac{s_i}{\sum_{i=1}^{n} s_i}$, and s_i is the i th singular value of $\mathbb{A}$ and $v_i = \frac{s_i}{\sum_{i=1}^{n} s_i}$ is the i th singular value of $\mathbb{B}$. Similar with Eq. 1, n is the number of singular values of $\mathbb{X}$.

Therefore, when the data mutual information between subsets of a dataset is higher, it indicates a stronger relationship between the datasets based on their singular values.

This stronger relationship often leads to improved classification performance, irrespective of the specific observations within the classes. The higher data mutual information signifies a more significant shared information content between the datasets, which can enhance the discriminative features extracted by classification algorithms, ultimately resulting in better classification performance.

4 Machine Learning Models

In this section, we describe the classifiers used in the following experiments briefly and the reasons why we choose them in imbalanced learning scenarios. The chosen classifiers fall into two main categories: classical machine learning models and deep learning models. The four learning models used include Support Vector Machines (SVM), Convolutional Neural Networks (CNN), Deep Neural Networks (DNN), Transformers, and its variant Vision Transformers (ViT).

4.1 SVM

SVMs are generally considered as a stable classifier for their ability to generalize well to unseen data, making them less prone to overfitting [15, 16].

SVCs short for Support Vector Classifiers are a popular choice for many machine learning applications due to their ability to achieve high accuracy while still being computationally efficient. The basic idea behind SVC is as follows.

Given training dataset $D = \{(x_i, y_i)\}_{i=1}^{N} (x^i \in R^d, y_i \in \{-1, +1\})$ where x^i represent a training sample with d features and y_i is the class label for each training sample. The goal of SVC is to solve the optimization problem of a dual problem expressed as Eq. 3.

$$max_\alpha \sum_{i=1}^{m} \alpha_i - \frac{1}{2} \sum_{i=1}^{m} \alpha_i \alpha_j y_i y_j K(x_i, x_j)$$

$$\text{subject to } 0 \le \alpha_i \le C, i = 1, 2, \ldots m, \sum_{i=1}^{m} \alpha_i y_i = 0 \tag{3}$$

where α_i is the regularization parameter used for balancing the importance of maximizing the margin between classes and minimizing the classification error on x_i. The $K(x_i, x_j)$ part denotes a kernel function and is used to map data into a higher dimensional space. The commonly-used Radial Basis Function (RBF) kernel is defined as $K(x_i, x_j) = exp(-\gamma \|x_i - x_j\|^2)$.

Once the optimization problem is solved, to predict the class label of a given sample x, the decision function $f(x) = sign(\sum_{i=1}^{m} \alpha_i y_i K(x_i, x_j) + b)$ is used.

Compared with other learning models, kernel SVMs are distinctive and more well-suited for tasks involving non-linear relationships, high-dimensional data, and robust classification requirements because of their strong abilities to learn complex decision boundaries and separate normal data points from outliers. Additionally, SVMs allow for fine-tuning of hyperparameters C (regularization parameter) and kernel parameters (γ in RBF Kernel) [17], which can be crucial for optimizing the performance of models in imbalanced learning.

4.2 CNN

CNNs are a class of DL networks and is popular in processing and analyzing visual data, such as images. A typical CNN architecture can be denoted by tuple $\{Conv^1, \sigma, MaxPool^1, Conv^2, \sigma, MaxPool^2, ...Flatten, FC^1, FC^2\}$, where $Conv^i$, $MaxPool^i$ and FC^i ($i = 1, 2...$) represent the i th convolutional layer, the i th maxpooling layer and the i th fully connected layer as well as σ represents the activation such as ReLU used. Deeper architectures incorporate more layers, which allows the CNN to capture intricate features and patterns in the input data.

The typical input of a CNN is an image which can be represented as a tensor $I \in \mathbb{R}^{H \times W \times C}$ where H is the height of the image, W is the width of the image and C is the number of the channels in the image. The core operation in CNNs is the convolution operation which is in essence to obtain the output feature map $O_{i,j,d}$ as is shown in Eq. 4 through applying a filter $K \in \mathbb{R}^{M \times N \times C \times D}$ to the input image I.

$$Q_{i,j,d} = \sigma\left(\sum_{m,n,c} I_{(i \times S+m)-P,(j \times S+n)-P,c} \times K_{m,n,c,d} + b_d\right) \qquad (4)$$

where $O_{i,j,d}$ represents the output activation at position (i, j) for the d th channel in the feature map, σ is the activation function, $I_{i \times S+m, j \times S+n, c}$ represents the input value at position $(i \times S + m, j \times S + n, c)$ in the input image, $K_{m,n,c,d}$ is the weight of the filter at position (m, n, c, d), b_d denotes the bias for the d th channel, S and P represent the amount of shift when sliding the filter as well as that of padding to the image I.

While CNNs may face challenges in handling imbalanced data distributions, their inherent ability to capture complicated patterns and distinguish between classes makes them valuable for various classification tasks [18, 19]. Especially, the 1D-CNNs are adapted for tabular data classification tasks and the classification for datasets stored in.csv format that are used in our experiments can benefit from 1D-CNNs.

4.3 DNN

DNNs are also typical learning models which consist of an input layer, multiple hidden layers, and an output layer and are commonly used in various classification tasks, particularly when dealing with moderate-sized datasets just as what used in our experiments [20].

The learning capabilities of a DNN depends on the sequential execution of some operations such as feedforward computation, loss function evaluation and backpropagation, facilitating the iterative refinement of the network parameters to enhance model performance and optimize the mapping from input data $x \in \mathbb{R}^n$ to the corresponding output prediction $\hat{y} \in \mathcal{Y}$ ($\mathcal{Y}$ represents the set or space that y belongs to, for example, for a binary classification $y \in \{0,1\}$).

The feedforward computation involves computing the pre-activation as is shown in Eq. 5 for each layer l(where $l = 1, 2, ...L$) of a DNN with L hidden layers.

$$z^{(l)} = W^{(l)}a^{(l-1)} + b^l \tag{5}$$

where $W^{(l)} \in R^{n_l \times n_{l-1}}$ is the weight matrix, $a^{(l-1)} \in R^{n_{l-1}}$ is the activation of the previous layer and $b^{(l)} \in R^{n_l}$ is the bias. Consequently, an activation function f is applied to $z^{(l)}$ to get the activation $a^l = f(z^{(l)})$ and the final result $\hat{y} = g(z^{(L)})$ is the predicted output of the DNN.

Compared to CNNs, which are more efficient for visual data such as images, DNNs offering versatility across different domains such as text, audio, and numerical data. Both of them face challenges when dealing with imbalanced datasets and need strategies such as data augmentation and resampling techniques to mitigate the imbalance and improve model performance on minority classes.

4.4 Transformer and ViT

The Transformer model is a powerful model for various natural language processing (NLP) tasks and the core of them is the self-attention mechanism depicted in Eq. 6. Additionally, the self-attention mechanism is extended to multi-head attention, which means the transformer models compute multiple attention mechanisms in parallel [21, 22].

$$\text{Attention}(\mathbf{X}, Q, K, V) = \text{softmax}(\frac{XW_Q(XW_K)^T}{\sqrt{D_k}}) \cdot \mathbf{X}_V \tag{6}$$

where $\mathbf{X} \in R^{N \times D}$ represents the input embedding matrix (N is the number of patches, D is the dimension of each patch embedding), W_Q, W_K and W_V are learnable linear projection matrices, D_k is the dimension of the key vectors.

The Vision Transformer (ViT) [23] is a popular model architecture that applies the Transformer architecture to image classification tasks. It breaks an image $X \in \mathbb{R}^{H \times W \times C}$ into N non-overlapping patches $p_j (j \in \{1, 2, ..., N\})$ which are then added to positional encoding vector $E_{pos_j} \in \mathbb{R}^D$, as is shown in Eq. 7.

$$z_j^{(0)} = W_e \cdot Flatten(p_j) + b_e + E_{pos_j} \forall j \in \{1, 2, ..., N\} \tag{7}$$

where $W_e \in \mathbb{R}^{D \times P^2 C}$ is a trainable weight matrix, D is the dimension of embedding space, P is the width and height of each image patch, C is the channel number of the image X.

By concatenating $z_j{}^{(0)}$ from Eq. 7 with a learnable classification token *token*, the initial embedding vector of the flattened patches after positional encoding $z^{(0)}$ shown in Eq. 8 is obtained.

$$\mathbf{z}^{(0)} = \{token, z_1^{(0)}, z_2^{(0)}, ..., z_N^{(0)}\} \in \mathbb{R}^{(M+1) \times D} \tag{8}$$

By passing each $\mathbf{z}^{(l)} l \in \{1,2, ..., L\}$ (L is the total number of Transformer encoder layers) to multiple layers of the Transformer encoder for the necessary processing including Multi-Head Self-Attention (MHSA) shown in Eq. 9, layer normalization, residual connections, and feed-forward networks, the final $\mathbf{z}^{(L)}$ can be obtained for the further classification.

$$head_h = \text{Attention}(QW_h^Q, KW_h^K, VW_h^V)$$

$$\text{MHSA}(Q, K, V) = \text{concat}(head_1, head_2, ...head_h)W^O \tag{9}$$

where W_h^Q, W_h^K, W_h^V are learned projection matrices, W^O denotes a learnable weight matrix and the attention mechanisms is same with Eq. 6.

While transformer and ViT models have stronger recognition capacities especially in computer vision and natural language processing, they have certain weaknesses and limitations inevitably. One primary drawback for everyday tasks is their high computational demands and the extensive training data needed, especially for imbalanced data, the minority class often lacks sufficient samples for effective model training. Additionally, both transformer and ViT models require input data to be flattened, restricting their capability to capture hierarchical features within the data.

5   Performance Evaluation Metrics

When faced with class imbalance, the traditional overall accuracy can only reflect the classification effect of the classifier as a whole, and cannot reflect the classification accuracy of the minority class [24]. Usually, the F1-score is used more because it is a harmonic mean of precision and recall as is shown in Eq. 10, providing a balance between the two metrics.

$$F1 - score = \frac{2 * precision * recall}{precision + recall} \tag{10}$$

Moreover, while the F1-score can provide valuable insights in performance evaluation, it still has limitations in offering a comprehensive assessment of learning performance and the potential for biased evaluations, particularly in imbalanced learning. Hence, we introduce a metric termed the d-index, as proposed by Han.*et al.* [25]. The d-index is defined by Eq. 11 and the larger the d-index value, the better the predictability of learning models.

$$d - index = \log_2(1 + a) + \log_2(1 + \frac{s + p}{2}) \tag{11}$$

where a, s, and p represent the corresponding accuracy, sensitivity and specificity, respectively.

6 Results

In this section, a number of experiments on different models and datasets are conducted and discussed. All the experiments are divided into two groups for validating the proposed indicators.

6.1 Dataset and Model Structure

In our experimental setup, we use four distinct types of models: CNN, DNN, SVM and Transformer (including its variant ViT). These models are utilized across four datasets, namely, CIFAR-10, Liver Cirrhosis, Airline Satisfaction and Body Smoking Signal, to perform classification tasks.

CIFAR-10, is a commonly-used image dataset for multiple classification tasks. The Liver Cirrhosis dataset contains 18 features and is used for multiple classification tasks related to liver health. The class labels are D(death), C(censored) and CL (censored due to liver transplantation). The Airline Satisfaction dataset with 21 features is utilized for binary classification tasks focusing on airline customer satisfaction. The class labels are Neutral or Dissatisfied and Satisfied. The Body Smoking Signal dataset featuring 24 dimensions is also used for binary classification tasks, particularly related to smoking signals in the body. The class labels are Non-smoking and Smoking.

Figure 1 shows the separability of datasets Liver Cirrhosis, Airline Satisfaction, and Body Smoking Signal through t-SNE plots. It is obvious that there are significant overlaps between classes of the three datasets to some extent, which reveals that the three datasets are non-linearly separable.

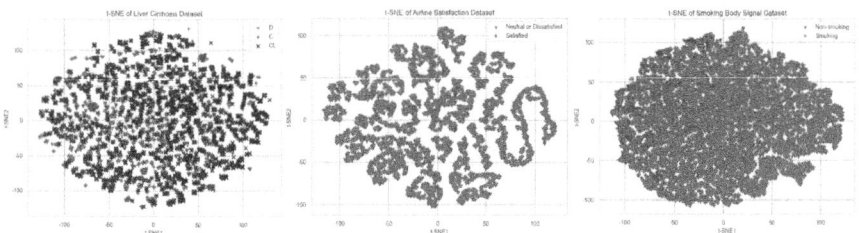

Fig. 1. t-SNE plots of three datasets Liver Cirrhosis, Airline Satisfaction, and Body Smoking Signal.

Based on the four datasets, the four types of models are customized, respectively. The CNN models are divided into two types: 1D-CNN and 2D-CNN (AlexNet). For Liver Cirrhosis, Airline Satisfaction, and Body Smoking Signal datasets, 1D-CNN models with different layers including Inputlayer, Conv1D, Activation, Batch Normalization, MaxPooling1D, Flatten, Dense, and Dropout are used. The CIFAR-10 dataset uses a 2D-CNN model (AlexNet). Similarly, DNN models with Inputlayer, Batch Normalization, Flatten, and Dense are customized for all datasets. Transformer architectures with MultiHeadAttention are employed, along with layers of Inputlayer, Batch Normalization, Flatten, and Dense. Notably, the ViT model for CIFAR-10 uses ViT-B16 as the base architecture. For more detailed information, please refer to our GitHub repository: https://github.com/nancy20240601/imbalanced-learning.

The SVM model for three datasets Liver Cirrhosis, Airline Satisfaction, and Body Smoking Signal utilizes the Radial Basis Function (RBF) kernel because of their non-linear separability reflected through Fig. 1. Because of limitations of computation capability in our experiments, we exclude the use of SVM for CIFAR-10 and the two primary parameters C and γ are customized through iteration on the possible lists of parameters ranges. Although the ranges of C and γ are between a small value (e.g., 0.001) to a large value (e.g., 10) or higher, higher values of C and γ will lead to higher overfitting risks. Therefore, we choose the ranges of C and γ are between 0.1 and 10.0. The detailed settings of parameters C and γ are outlined in Table 1.

Table 1. Settings of parameters C and γ of SVMs.

Model	Parameter C	Parameter γ
Liver Cirrhosis	8.1	0.1
Airline Satisfaction	1.1	0.1
Body Smoking Signal	0.1	5.1

6.2   Case Study

In our case study, we perform two groups of experiments to verify that the data entropy and data mutual information are two key factors that bring impacts on classification performance in imbalanced learning.

Group one is used to verify the classification performance may show bias even when the observations of different classes are evenly distributed. The data entropy can quantify the information of classes in a dataset effectively. On the contrary, Group two is used to demonstrate that rebalancing the data entropy of each class and increasing data entropy of each class can improve the classification performance to some extent.

Group One: Quantifying Information of Classes by Data Entropy

In this group of experiments, we firstly apply Eq. 1 to compute the data entropy (as illustrated in Table 2) of each class of four datasets used. Despite having an equal number of observations for each class (7000 for Liver Cirrhosis, 9000 for Airline Satisfaction, 19000 for Body Smoking Signal, and 4000 for CIFAR-10 in the training set), the data entropy of each class is not equal, which reflects the information imbalance of each class.

Furthermore, we calculate the data mutual information between different classes of the four datasets used. The data mutual information of two classes of Airline Satisfaction dataset and Body Smoking Signal dataset are 0.9982 and 4.3680, respectively. The data mutual information between two classes, that is, D and C, D and CL as well as C and CL of Liver Cirrhosis dataset are 1.6362, 1.5926 and 1.6446. For CIFAR-10 dataset, there are 45 data mutual information between every pair of classes. The data mutual information between Class Frog and other classes achieves the maximum 16.6851 while the data mutual information of Cat Class between other classes achieves the minimum 16.3278.

Table 2. Data entropy of each class of four datasets.

Dataset	Observations	Class Label	Data Entropy
Airline Satisfaction	9,000	Neutral or Dissatisfied	0.6184
		Satisfied	0.3977
Body Smoking Signal	19,000	Non-smoking	2.4386
		Smoking	2.4549
CIFAR-10	4,000	Airplane	8.7527
		Automobile	9.1822
		Bird	8.9730
		Cat	8.9334
		Deer	9.0935
		Dog	8.9415
		Frog	9.2638
		Horse	9.0466
		Ship	8.7210
		Truck	9.1268
Liver Cirrhosis dataset	7,000	D	0.8413
		C	0.9031
		CL	0.8521

Note: For the Liver Cirrhosis dataset, D, C and CL denote the three-class label death, censored and censored due to liver transplantation, respectively

The following Fig. 2 presents the t-SNE plots of the three datasets Liver Cirrhosis dataset, Airline Satisfaction, their classification performance on CNN models, providing a visual representation of different information richness embedded in different datasets which can be extracted to benefit the classification tasks.

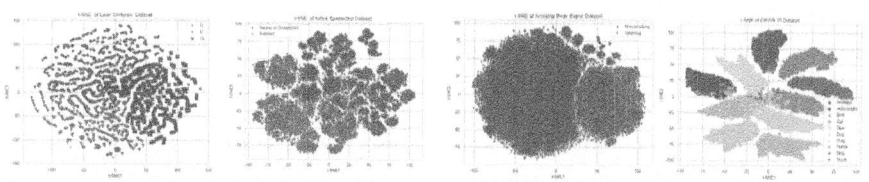

Fig. 2. t-SNE plots of four datasets used on CNN models. From left to right are the Liver Cirrhosis dataset, Airline Satisfaction dataset, Body Smoking Signal dataset and CIFAR-10 dataset.

The following Fig. 3 presents the detailed results of experiments on the four datasets across four models. For the binary classification tasks on Airline Satisfaction dataset and Body Smoking Signal dataset, it is observed that the class with larger data entropy archives lower F1 scores and d-index values across all models used. This observation confirms that the reason contributing to bias in classification is from the imbalance of information of classes rather than only that of the observations. Higher data entropy represents more information in the corresponding class and bring challenge for classification.

In the experimental results of the Liver Cirrhosis dataset, it is observed that the class C with highest data entropy always achieves lowest F1 scores and d-index values in most cases. On the contrary, the class D and the class CL with lower data entropy achieve higher F1 scores and d-index values, which indicates reduced randomness or information within the class in turn benefits classification performance, confirming with our insights provided by Eq. 1 again.

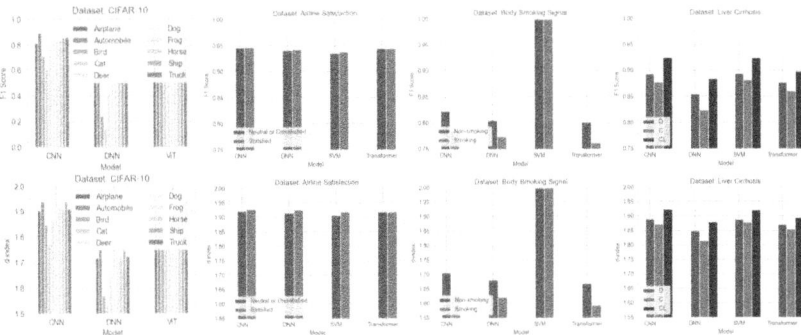

Fig. 3. F1 scores and d-index values of four datasets on different models. Due to limitations in computational resources during our experiments, we have excluded the results of the SVM model on the CIFAR-10 dataset.

However, there are exceptions to this trend in some results. For the class D and the class CL which have similar lower data entropy, we observe that the class CL with comparably higher data entropy achieves higher F1 scores and d-index values. The reason behind this observation is that the data mutual information (1.6446) between the class C and the class CL is a little higher than that (1.6362) between the class D and the class C.

Due to the higher dimensionality of the CIFAR-10 dataset, the complexity of its results is increased. The complexity may come from different factors such as the learning ability of models, characters of dataset themselves and relationships between classes. In our analysis, we combine the data entropy of each class and data mutual information together to comprehensively examine the classification bias present in the dataset.

Based on Table 2 and Fig. 3, it is observed that the Ship Class with the lowest data entropy achieves the second-highest F1 score and d-index value across all the three models, which proves that the classes with low data entropy have low information and are advantageous for learning algorithms or classifiers to classify instances belonging to these classes more accurately than those classes with higher data entropies.

The Cat Class achieves the lowest F1 scores and d-index values on all the three models. This could be attributed to its higher data entropy but lower data mutual information (an average value of 16.3278), which is the minimum average data mutual information between two classes of CIAR-10.

The average data mutual information between Cat Class and other nine classes are 16.2829, 16.3162, 16.3919, 16.4367, 16.1416, 16.5559, 16.2686, 16.2027 and 16.3474, respectively.

The above experimental results highlight the significance of data entropy and data mutual information as important factors that can affect bias in classification performance, even when observations in each class are balanced. In general, higher data entropy indicates more information within a class, which can impede classification performance. On the other hand, higher data mutual information facilitates classification by enhancing the shared information between classes.

Group Two: Better Classification Performance Through Balancing Data Entropy
In this group of experiments, our goal is to demonstrate that better classification performance can be attained by balancing the data entropy of classes.

The following experiments are carried on the Liver Cirrhosis dataset in three cases. Case 0 is used as a benchmark and keep the original data entropies of classes. In Case 1 and Case 2, we use oversampling (SMOTE) to change the data entropy of each class and try to reduce the data entropy difference between classes. The detailed data entropy and difference of classes before and after resampling is show in Table 3 and Table 4.

Table 3. Data entropy of classes in three cases for the Liver Cirrhosis dataset.

Class Label	Case 0	Case 1	Case 2
D	0.8413	0.8446	0.8339
C	0.9031	0.9035	0.8944
CL	0.8521	0.8530	0.8474

Table 4. Data entropy difference of classes in three cases for the Liver Cirrhosis dataset.

Case	D ~ C	D ~ CL	C ~ CL
Case 0	0.0618	0.0510	0.0108
Case 1	0.0589	0.0506	0.0083
Case 2	0.0605	0.0470	0.0135

In Fig. 4, we observe a slight increase in the data entropies of the three classes from Case 0 to Case 1. This increase, together with a smaller difference between classes, results increase on the d-index values across the classes. Typically, the class D with the lowest data entropy achieves some improvements on all the four models. And in Case 1, the bias in classification on three classes are indeed reduced, proving balancing the data entropy of classes can reduce classification bias and if increasing the data entropy of each class at the same time can improve classification performance. The increase of data entropy means increased data variability of classes, which can provide the classifier with more information and potentially more discriminative features to distinguish between the classes.

In contrast, from Case 0 to Case 2, the more balanced data entropy between classes reduces the difference of classification to some extent, but due to the overall reduction in the data entropy of each class, the classification performance, as measured by the d-index, is lower compared to the benchmark Case 0.

Conversely, Case 1 shows an increase not only in data entropy but also in the data mutual information between every pair of classes. The higher values of data mutual information (1.6387, 1.5957, and 1.6456) in Case 1 also contribute to improving classification performance. On the other hand, in Case 2, the data mutual information between each pair of classes is decreased to values of 1.6213, 1.5807, and 1.6317. This reduction poses a disadvantage for classification tasks, indicating that a decrease in data mutual information hinders classification performance.

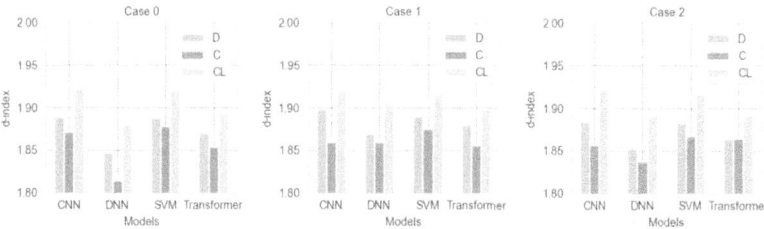

Fig. 4. Balancing the data entropy difference of classes of the Liver Cirrhosis dataset. Case 0 is used as a benchmark.

7 Conclusion

In this study, we propose the viewpoint that the deeper reason of classification bias lies in the information imbalance of classes in imbalanced learning. We propose an indicator data entropy based on singular values from SVD to reflect the information imbalance between classes of a dataset. In addition, we propose the data mutual information, which effectively explain the complex information relationships among classes and the changes of data mutual information will bring impacts on classification performance. Through a series of diverse experiments conducted on four datasets across four models, we validate the efficacy of our proposed indicators. Our experimental results prove that the data entropy and data mutual information are two important factor posing impacts on classification performance. Minimizing the difference between data entropy of classes and increasing the data entropy and data mutual information simultaneously can improve classification performance and reduce classification bias among classes.

While we achieve insights from previous work, more improvements need to be made in our future study. For example, the current definition of data mutual information focuses only on the information exchange between two classes. In complex multiple classification tasks, as the number of classes grows, a single class may exhibit complex relationships with all other classes in the dataset. This complexity introduces additional challenges in computing data mutual information accurately, emphasizing the need for a more comprehensive approach to capture the intricate interdependencies among all classes.

Additionally, in the future work, we hope to explore effective strategies, including but not limited to resampling techniques, for the purpose of adjusting data entropy and data mutual information of datasets so that the classification performance can be improved more significantly in imbalanced learning.

Acknowledgments. The authors acknowledge National Natural Science Foundation of China (Grant: 62066039), Natural Science Foundation of Qinghai Province (Grant: 2022-ZJ-925), and the "111" Project (D20035).

References

1. Rezvani, S., Wang, X.: A broad review on class imbalance learning techniques. Appl. Soft Comput. **143**, 110415 (2023)
2. Zhu, R., Guo, Y., et al.: Adjusting the imbalance ratio by the dimensionality of imbalanced data. Pattern Recogn. Lett. **133**, 217–223 (2020)
3. Fajardo, V.A., Findlay, D., et al.: On oversampling imbalanced data with deep conditional generative models. Expert Syst. Appl. **169**, 114463 (2021)
4. Bai, L., Ju, T., et al.: Two-step ensemble under-sampling algorithm for massive imbalanced data classification. Inf. Sci. **665**, 120351 (2024)
5. Guan, S., Zhao, X., et al.: AWGAN: an adaptive weighting GAN approach for oversampling imbalanced datasets. Inf. Sci. **663**, 120311 (2024)
6. Sinha, G.R., Suri, J.S.: Cognitive Informatics, Computer Modeling, and Cognitive Science, 1st edn. Academic Press, New York (2020)
7. Alter, O., et al.: Singular value decomposition for genome-wide expression data processing and modeling. Proc. Natl. Acad. Sci. **97**(18), 10101–10106 (2000)
8. Han, H., Teng, J., Xia, J., Wang, Y., Guo, Z., Li, D.: Predict high-frequency trading marker via manifold learning. Knowl.-Based Syst. **213**, 106662 (2021)
9. Cincotta, P.M., Giordano, C.M., et al.: The Shannon entropy: an efficient indicator of dynamical stability. Physica D Nonlinear Phenom. **417**, 132816 (2021)
10. Saraiva, P.: On Shannon entropy and its applications. Kuwait J. Sci. **50**, 194–199 (2023)
11. Ji, X., Peng, S., Yang, S.: Imbalanced binary classification under distribution uncertainty. Inf. Sci. **621**, 156–171 (2023)
12. Fu, Y., Shang, C., et al.: ECMEE: expert constrained multi-expert ensembles with category entropy minimization for long-tailed visual recognition. Neurocomputing **576**, 127357 (2024)
13. Guermazi, R., Chaabane, I., et al.: AECID: asymmetric entropy for classifying imbalanced data. Inf. Sci. **467**, 373–397 (2018)
14. Yang, Z., Qi, W., et al.: MIB-Net: balance the mutual information flow in deep learning network for multi-dimensional segmentation of COVID-19 CT images. Biomed. Signal Process. Control **95**, 06376 (2024)

15. Yang, C., Yang, J., et al.: Margin calibration in SVM class-imbalanced learning. Neurocomputing **73**, 397–411 (2009)
16. Cai, W., Cai, M., et al.: Three-way imbalanced learning based on fuzzy twin SVM. Appl. Soft Comput. **150**, 111066 (2024)
17. Zhang, M., Treder, M., et al.: Explaining the predictions of kernel SVM models for neuroimaging data analysis. Expert Syst. Appl. **251**, 123993 (2024)
18. Zhao, X., Zhao, X., et al.: A balanced random learning strategy for CNN based Landsat image segmentation under imbalanced and noisy labels. Pattern Recognit. **144**, 109824 (2023)
19. Taherkhani, A., Cosma, G., et al.: AdaBoost-CNN: an adaptive boosting algorithm for convolutional neural networks to classify multi-class imbalanced datasets using transfer learning. Neurocomputing **404**, 351–366 (2020)
20. Pawan, S.J., Rajan, J.: Capsule networks for image classification: a review. Neurocomputing **509**, 102–120 (2022)
21. Lin, T., Wang, Y., et al.: A survey of transformers. AI Open **3**, 111–132 (2022)
22. Zeng, C., Zhang, J., et al.: Multiple attention mechanisms-driven component fault location in optical networks with network-wide monitoring data. J. Opt. Commun. Network. **15**, 9–19 (2023)
23. Thisanke, H., Deshan, C., et al.: Semantic segmentation using vision transformers: a survey. Eng. Appl. Artif. Intell. **126**, 106669 (2023)
24. Naidu, G., Zuva, T., et al.: A review of evaluation metrics in machine learning algorithms. In: Silhavy, R., Silhavy, P. (eds.) CSOC 2023, pp. 15–25. Springer, Heidelberg (2010). https://doi.org/10.1007/978-3-031-35314-7_2
25. Han, H., Wu, Y., et al.: Interpretable machine learning assessment. Neurocomputing **561**, 126891 (2023)

Analysis of the Changes in Microbial Community During the Fermentation of Feng-Flavour Baijiu

Yongli Zhang[1,2], Gang Xing[2], Zirui Zhang[2], and Yaodong Chen[1(✉)]

[1] College of Life Sciences, Northwestern University, Xi'an 710069, Shaanxi, China
ydchen@nwu.edu.cn

[2] Shaanxi Xifeng Liquor Co., Ltd., Baoji 721400, Shaanxi, China

Abstract. Feng-flavour Baijiu is one of the four famous liquors in China. This study analyzed the changes of microbial community structure in the fermented grains of Feng-flavour Baijiu. A total of 133 fungi and 688 bacteria were detected at the genus level. The dominant fungi were *Pichia, Naumovozyma, Aspergillus, Thermoascus* and *Lichtemia,* while the dominant bacteria are *Lactobacillus, Streptomyces, Kroppenstedtia, Bacillus, Aquabacterium, Staphylococcus, Weissella, Pediococcus* and *Saccharopolyspora.* The composition of the dominant microorganism in the fermented grains of Feng-flavour Baijiu was different from other flavor Baijiu. It had unique characteristics, which was one of the important factors in the formation of Feng-flavour Baijiu.

Keywords: Feng-flavour Baijiu · Microbial community · Core function microorganisms

1 Introduction

Chinese liquor (Baijiu) brewing has a long history in China. Traditional Baijiu is mainly distilled liquor [1], which uses high starch raw materials as the fermentation substrate, distiller's yeast as the saccharifying agent, and solid, semi-solid or liquid fermentation method is used for distillation, aging and blending into alcoholic beverages, and is listed as one of the world's famous distilled liquors [2]. Liquor has regional uniqueness. Because of the differences in fermentation substrates, types of distiller's yeast, fermentation technology and natural geographical environment, Baijiu with different characteristics and flavor types was finally formed. Feng-flavour Baijiu is one of the four traditional famous liquors in China, it is fermented in the soil cellar, and the pit mud is renewed once a year. In a year's production cycle, it can be divided into six production stages. The fermentation period of each stage is 30 days, and the produced Baijiu is called base Baijiu. The base Baijiu will be stored in a large conservator for more than 3 years. After blending, it has formed the typical style characteristics of Feng-flavour Baijiu, which is mellow and elegant, sweet and refreshing, harmonious in various flavors, durable and clean.

H. Han (Ed.): SDSC 2024, CCIS 2158, pp. 110–119, 2024.
https://doi.org/10.1007/978-3-031-67871-4_8

The formation of the flavor of Chinese Baijiu is mainly due to the compounds formed by the growth and metabolism of the various, numerous and interacting microorganisms. The fermentation microorganisms of Baijiu mainly include bacteria, yeast and mold, among which bacteria can produce many flavor metabolites [3]. Yeast is the key microorganism [4, 5] for ethanol production, organic acid production and the formation of a variety of important ester flavor substances. A variety of enzymes secreted by mold can consume raw materials [6]. These microorganisms are distributed throughout the whole fermentation system, and various microorganisms interact to form a unique brewing microbial system, which can also be called Baijiu brewing microecology. Different flavor Baijiu may have unique flavor characteristics due to the difference microecological components in the brewing process [7]. High throughput sequencing (HTS) is one of the commonly used methods to study the microbial diversity of fermented food, and has been widely used in the research of microbial community structure in the production of fermented food, such as Korean kimchi [8], Chinese fen-flavor Baijiu solid fermentation microbial community [9], cheese [10], sake [11], sour dough [12], Kefir [13], wine [14], etc. It is also used to analyze the fermentation production of Chinese Baijiu with different flavors and it is suggested that the types and changes of different microbiota have a strong correlation with the different characteristics of Baijiu. In the production of Chinese Fen-flavour Baijiu, a study suggested that *Acetobacter* and *Lactobacillus* were the dominant bacteria, while *Candida* and *Pichia* were the dominant fungi [15]. And another study suggested that bacteria are mainly *Lactobacilliaceae*, fungi are mainly *Saccharomyceteceae* and *Saccharomycopsidaceae* [16]. In the first to seventh rounds of Maotai-flavour Baijiu stacking fermentation, it was suggested that the characteristic fungi were *Davidielaceae, Candida* and *Saccharomyces*, while the characteristic bacteria were *Paenibacillus, Flavobacterium, Delftia* and *Sphingomonas* [17]. Another study suggested that in the fermentation process of Maotai-flavor Baijiu, the main metabolic role in the fermentation process was *Pichia, Schizosaccharomyces,* and *Conjugate* [4]. In our previous study of the fermentation process of Feng-flavor Baijiu, showing that *Naumovozyma* and *Lactobacillus* are the absolute dominant microorganisms [18].

In this paper, high-throughput sequencing technology was used to analyze the microbial community structure and diversity during the brewing process of Feng-flavour Baijiu, and data mining was carried out on the dynamic changes and succession laws of microbial community in combination with bioinformatics analysis and multivariate statistical analysis, in order to find the dominant microbial species in the fermentation process of Feng-flavour Baijiu.

2   Materials and Methods

2.1 Sample Collection and Processing

Three pits with normal fermentation were randomly selected, and each pit was sampled 7 times according to the fermentation time of 0, 4, 9, 14, 18, 23 and 28 days, and each time point was sampled according to the "nine points" method shown in Fig. 1.

2.2 High Throughput Sequencing Analysis of Microorganisms in Fermented Grains

Extraction of Total DNA from Microbiota

DNA kit (Omega Bio-Tek, USA) and Nanodrop were used to quantify DNA and the quality of DNA was detected by 1.2% agarose gel electrophoresis. The extracted DNA from sample was stored at −80 °C for amplicon sequencing.

Amplicon Sequencing

The fungal region of ITS_ V1 was sequenced using primers ITS5F (GGAAG-TAAAAGTCGTAACAAGG) and ITS2R (GCTGCGTTCTTCATCGATGC), and the bacterial region of 16S v3-v4 was sequenced using primers F (ACTCCTACGGGAG-GCAGCA) and R (GGACTACHVGGGTWTCTAAT). After amplification, the purified 16S rRNA gene and ITS1 sequences were sequenced by Illlumina MiSeq platform, respectively, at BioNovoGene Co., Ltd. (Suzhou, China).

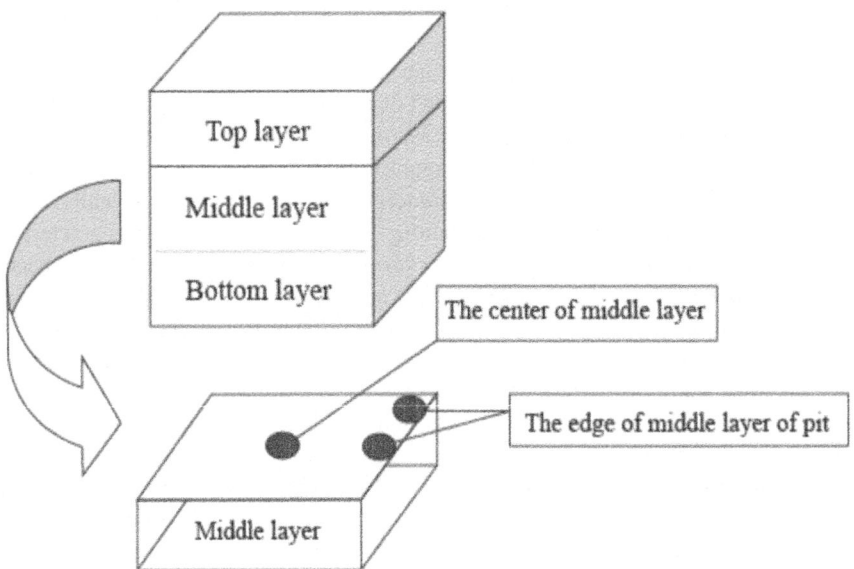

Fig. 1. Sampling method of fermented grains. The fermented grains in the pit are divided into upper, middle and lower layers. Three points are taken from the middle and the edge of each layer, and the samples from nine points are mixed evenly as the fermented grains samples of the pit.

2.3 Traceability Analysis of Brewing Microorganisms

According to the microbial population structure of fermented grains and fermentation environment, this study uses SourceTracker software to analyze the source of microorganisms in fermented grains and sets the microbial population of Daqu and environmental samples as the source, The microbial population of Feng-flavour fermented grains is the receiving end, running 1000 times, and other parameters are default.

3   Results

3.1   Community Alpha Diversity Analysis

Alpha diversity is mainly used to study community diversity within a certain lifetime. By evaluating a series of Alpha diversity indices, information such as species richness and diversity in environmental communities can be obtained. The commonly used index algorithm for microbial Alpha diversity is as shown in Eq. 1.

Chao: It is an index used to estimate the number of species (such as ASV) in a sample using the Chao1 algorithm. Chao1 is commonly used in ecology to estimate the total number of species, and the calculation formula is as follows. Among them, Schao1: estimated number of species. Sobs: The actual number of observed species. n1: Number of species containing only one sequence. n2: The number of species containing only two sequences.

$$S_{chao1} = S_{obs} + \frac{n_1(n_1 - 1)}{2(n_2 + 1)} \tag{1}$$

Coverage: refers to the coverage of each sample library, as shown in Eq. 2. The higher the value, the higher the probability that the sequence in the sample will be measured, while the probability of not being measured will be lower. This index reflects whether the sequencing results represent the true situation of microorganisms in the sample. Among them, n_1: the number of species containing only one sequence. N: The total number of sequences appearing in the sampling.

$$C = 1 - \frac{n_1}{N} \tag{2}$$

Pielou_e: used to estimate the community evenness index, as shown in Eq. 3. The larger value indicates a more uniform distribution of the community. H: It is the Shannon exponent calculated above, ln: logarithm with e as the base.

$$J = \frac{H}{\ln(S)} \tag{3}$$

Shannon: One of the indices used to estimate microbial diversity in a sample, as shown in Equation 4. It is similar to the Simpson diversity index and is commonly used to reflect community alpha diversity. The larger the Shannon value, the higher the community diversity. Among them, S_{obs}: the actual number of observed species, n_i: The number of sequences contained in the i-th species, N: All sequence numbers.

$$H_{shannon} = -\sum_{i=1}^{S_{obs}} \frac{n_i}{N} \ln \frac{n_i}{N} \tag{4}$$

Simpson: One of the indices used to estimate microbial diversity in a sample, as shown in Eq. 5, often used in ecology to quantitatively describe the biodiversity of a region. The larger the Simpson index value, the lower the community diversity. Among them, S_{obs}: the actual number of observed species, n_i: The number of sequences contained in the i-th species, N: All sequence numbers.

$$D_{simpson} = \frac{\sum_{i=1}^{S_{obs}} n_i(n_i - 1)}{N(N - 1)} \tag{5}$$

From Tables 1 and 2, it can be seen that the richness and diversity of fungi ultimately increase with fermentation, while bacteria show a decreasing trend, indicating that suitable microorganisms were selectively selected in the high acid and high alcohol environment of the mash.

Table 1. Alpha Diversity of Fungal Communities

Sample	Chao1 (%)	Goods coverage (%)	Pielou_e (%)	Shannon (%)	Simpson (%)
A0d	39.75 ± 4.17	0.99 ± 1.05E-05	0.38 ± 0.09	2.04 ± 0.22	0.65 ± 0.14
A8d	43.53 ± 8.64	0.99 ± 5.05E-05	0.39 ± 0.09	2.12 ± 0.14	0.69 ± 0.12
A14d	35.86 ± 2.50	0.99 ± 9.87E-05	0.36 ± 0.05	1.88 ± 0.06	0.63 ± 0.04
A20d	36.39 ± 6.09	0.99 ± 2.43E-05	0.34 ± 0.07	1.75 ± 0.04	0.61 ± 0.07
A27d	47.25 ± 6.96	0.99 ± 3.33E-05	0.42 ± 0.10	2.34 ± 0.05	0.71 ± 0.02

Table 2. Alpha diversity of bacterial communities

Sample	Chao1 (%)	Goods coverage (%)	Pielou_e (%)	Shannon (%)	Simpson (%)
A0d	1019.11 ± 52.97	0.99 ± 1.34E-02%	0.58 ± 0.04%	5.81 ± 0.35%	0.92 ± 0.02%
A8d	303.42 ± 26.01%	0.99 ± 1.46E-02%	0.22 ± 0.06%	1.75 ± 0.38%	0.53 ± 0.01%
A14d	153.64 ± 9.17%	0.99 ± 2.51E-05%	0.21 ± 0.02%	1.49 ± 0.17%	0.52 ± 0.02%
A20d	151.46 ± 17.35%	0.99 ± 6.50E-05%	0.25 ± 0.03%	1.81 ± 0.28%	0.56 ± 0.03%
A27d	294.21 ± 57.35%	0.99 ± 1.34E-03%	0.32 ± 0.03%	2.5 ± 0.16%	0.64 ± 0.05%

3.2 Diversity and Succession of Fungal Community in Fermented Grains

High throughput sequencing showed that there were 8 phyla of fungi in fermented grains, among which *Ascomycota* played a dominant role in all stages of the fermented grain samples (Fig. 2A). A total of 133 fungal genera were detected at the genus level. There were 5 dominant fungi, *Pichia, Naumovozyma, Aspergillus, Thermoascus* and *Lichteimia* (the average ASV of the samples was more than 1%). *Pichia* was dominant in the whole fermentation process, with an average relative abundance of 49% ± 2.88%, the relative abundant of *Naumovozyma, Aspergillus*, and *Thermoascus* in the whole process was relatively stable, the relative abundance of *Lichteimia* was low (Fig. 2B). It can be seen from Fig. 2C that the functions of fungi were mainly biosynthesis, production of precursor metabolites and energy, with respiration being the most prominent.

3.3 Bacterial Diversity and Population Structure Succession Law in Fermented Grains

High throughput sequencing showed that there were 37 bacterial populations in all fermented grains, among which *Firmicutes* played a dominant role in all fermented grains (Fig. 3A). During the fermentation, 688 bacterial genera were detected at the genus level, of which 9 genera were dominant (the average ASV in the sample was greater than 1%). Among them, *Lactobacillus* was dominant in the whole fermentation process, with an average relative abundance of 68.81 ± 9.07%, *Streptomyces* average relative abundance of 6.28 ± 7.30%, *Kroppenstedtia* and *Bacillus* average relative abundance of about 3.5%, *Aquabacterium, Staphylococcus, Weissella, Pediococcus* and *Saccharopolyspora*

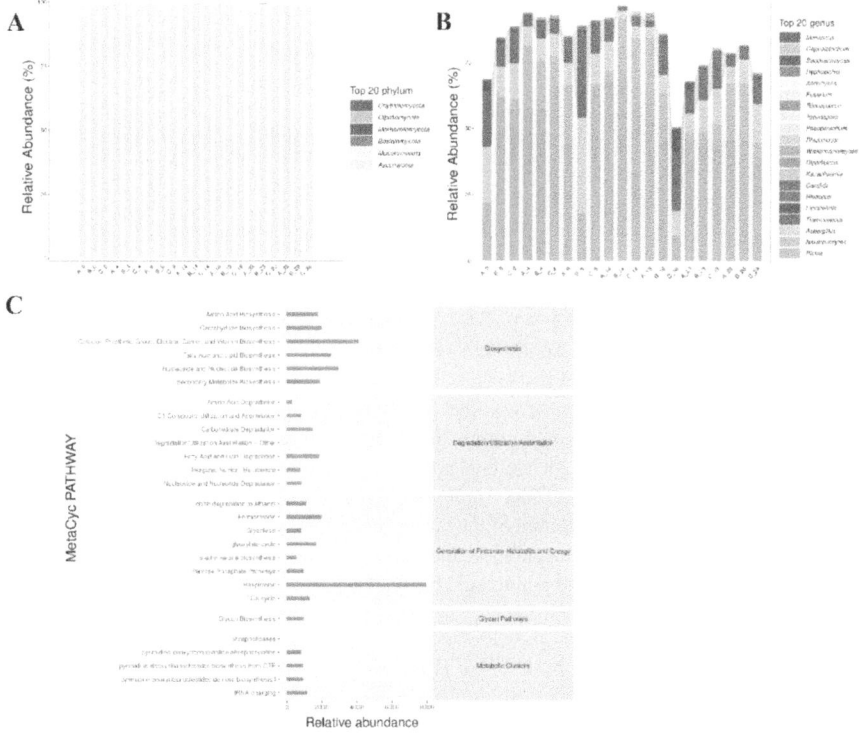

Fig. 2. Diversity and Succession Rules of Fungal Communities. (A) Histogram of Species Composition at the Level of Sect, there were 8 phyla of fungal populations in fermented grains, including *Ascomycota, Olpidiomycota, Mortierellomycota, Mucoromycota, Basidiomycota* unclassified_ Fungi and unidentified, among which, *Ascomycota* was the dominant phyla. (B) Horizontal species composition histogram in fermented grains at all stages, 133 fungal genera were detected, and the dominant fungi were 5 genera, namely *Pichia, Naumovozyma, Aspergillus, Thermoascus, Lichteimia.* (C) The prediction of functional potential of fungal community in fermented grains showed that the functions of fungi were mainly biosynthesis, production of precursor metabolites and energy.

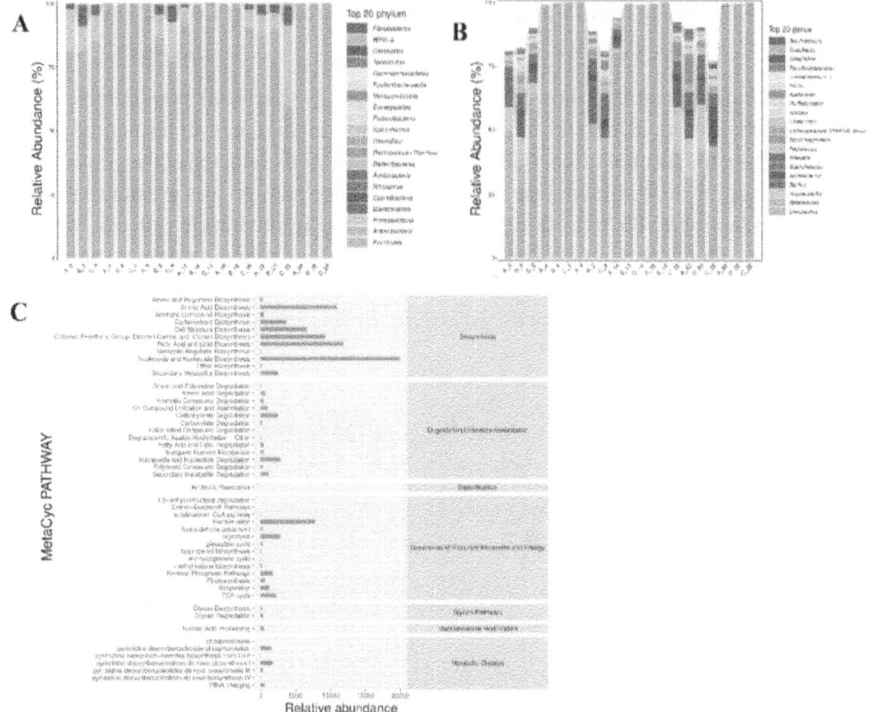

Fig. 3. The succession rule of bacterial diversity and population structure in fermented grains. (A) Horizontal species composition histogram, there were 37 phyla of bacterial populations in fermented grains, of which *Firmicutes* was the dominant phyla. (B) Genus horizontal species composition histogram in all fermented grains. 688 bacterial genera were detected at the genus level, of which 9 genera were dominant bacterial populations, namely *Lactobacillus, Streptomyces, Kroppenstedtia, Bacillus, Aquabacterium, Staphylococcus, Weissella, Pediococcus* and *Saccharopolyspora*. (C) The functional potential of the bacterial community in the fermented grains was predicted. The main function was biosynthesis, including nucleoside and nucleotide biosynthesis, fatty acid and lipid biosynthesis, and amino acid biosynthesis.

average relative abundance of about 1.4% (Fig. 3B). It can be seen from Fig. 3C that the main function of bacteria was biosynthesis, in which nucleoside and nucleotide biosynthesis, fatty acid and lipid biosynthesis, and amino acid biosynthesis were the main ones.

3.4   Traceability Analysis of Microorganisms in Fermented Grains

Figure 4 shows the microbial community structure of the fermented grains at different days, Daqu and the relevant environment. *Pichia, Thermoascus, Aspergillus* are the main fungi in Daqu and *Philemoniopsis* and *Pichia* are main fungi in grains and the environment. As the fermentation proceeds, the microbial community continues to adapt and eliminate, *Pichia, Thermoascus* and *Aspergillus* become the dominant fungus in the fermented grains. In addition, the fermented grains at 30 days also contains Rhizopus and Naumovozyma which may come from environment. It shows that Daqu has a great contribution to the fungus in fermented grains, and the ground and tools have also contributed a part of the fungus (Fig. 4A).

With the fermentation process, *Lactobacillus* becomes the absolute dominant bacteria in fermented grains. It shows that the environment and pit mud have a great contribution to the bacteria in fermented grains (Fig. 4B).

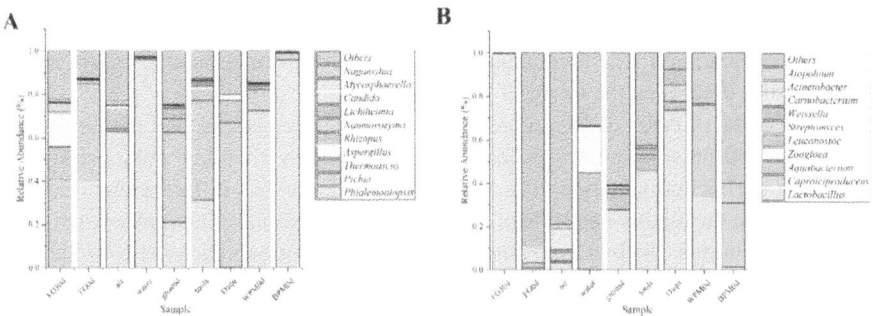

Fig. 4. The dominant microorganisms in all samples (A) The dominant fungi in all samples are mainly *Philemoniopsis* and *Pichia*. (B) The dominant bacteria in all samples are mostly *Lactobacillus*.

4 Discussion and Conclusion

Feng-flavour Baijiu was one of the four famous liquors in China. After comparison, the composition of the dominant bacteria in the fermented grains of Feng-flavour Baijiu was different from that of the three flavor types of Maotai-flavour Baijiu, Fen-flavour Baijiu, and Louzhou-flavour Baijiu. The dominant fungal strains in Feng-flavour Baijiu fermented grains were Pichia, Naumovozyma, Aspergillus, Thermoascus, and Lichtemia. The dominant bacteria were Lactobacillus, Streptomyces, Kroppenstedtia, Bacillus, Aquabacterium, Staphylococcus, Weissella, Pediococcus, and Saccharopolyspora. Pichia, Naumovozyma, and Lactobacillus were the absolute dominant microorganisms throughout the fermentation process.

The dominant fungi in Maotai-flavour Baijiu fermented grains were Alternaria, Ciliophora, Cyphellophora, Botryotinia, Issacchenkia, and Pichia, while the dominant bacteria included Escherichia-Shigella, Lactobacillus, Clostridium, Streptococcus, Actinobacillus, Bifidobacterium, Peptoclostridium, and Citrobacter [19]. In Fen-flavour Baijiu, the dominant fungi were Pichia, Candida, Kazachstania, Saccharomycosis, Saccharomyces, Wickerhamomyces, and Aspergillus, and the dominant bacteria were Lactobacillus, Bacillus, Pseudomonas, Kroppenstedtia, Weissella, and Acinetobacter [20]. For Luzhou-flavour Baijiu, the dominant fungi were Kazachstania, Aspergillus, Thermoyces, Thermoascus, and Eurotium, and the dominant bacteria were Lactobacillus, Bacillus, Weissella, Dysgonomonas, Comamonas, and Ruminococcus [21]. Certain fungi

(Naumovozyma, Lichteimia) and bacteria (Aquabacterium, Staphylococcus, Pediococcus, and Saccharopolyspora) in Feng-flavour Baijiu were not the dominant strains in other flavor types, while Pichia, Aspergillus, Thermoascus, Lactobacillus, Streptomyces, Kroppenstedtia, Bacillus, and Weissella were.

Baijiu fermentation is a natural and open process, and the fermentation of grains involves a large number of microorganisms from raw materials, tools, ground, air, and other sources. Recently, the concept of Baijiu production areas has highlighted the importance of environmental microorganisms in Baijiu brewing. The microbial traceability analysis of Feng-flavour Baijiu shows that the fungi mainly come from Daqu, and the bacteria mainly come from pit mud and the environment. This indicates that Daqu, as a starter for fermentation, provides abundant microorganisms in fermented grains, and its characteristics determine the flavor of Baijiu to a certain extent.

Different flavor types of Baijiu are produced in different areas, and the use of various strains and brewing processes leads to differences in microbial composition and succession during fermentation. The final distilled liquor has different trace components, affecting the quality and style of the liquor. The microbial community structure of Feng-flavour Baijiu is distinct from other types and has unique characteristics, which significantly contribute to the formation of Feng-flavour Baijiu. Future research will involve in-depth mechanistic analysis of brewing microorganisms' fermentation through transcriptomics and metabolomics, commonly used in industrial microbiology and human health research [22, 23].

References

1. Zhao, S., Yang, C., Dou, C., Xu, M., Liao, Y.: Research progress of brewing microorganisms in Baijiu production. China Brewing **31**, 5–10 (2012)
2. Wang, H., Zhang, X., Zhao, L., Xu, Y.: Analysis and comparison of the bacterial community in fermented grains during the fermentation for two different styles of Chinese liquor. J. Ind. Microbiol. Biotechnol. **35**, 603–612 (2008)
3. Zhang, R., Wu, Q., Xu, Y.: Aroma characteristics of Moutai-flavour liquor produced with Bacillus licheniformis by solid-state fermentation. Lett. Appl. Microbiol. **57**, 11–18 (2013)
4. Song, Z., Du, H., Zhang, Y., Xu, Y.: Unraveling core functional microbiota in traditional solid-state fermentation by high-throughput amplicons and metatranscriptomics sequencing. Front. Microbiol. **8**, 1294–1307 (2017)
5. Wu, Q., Chen, B., Xu, Y.: Regulating yeast flavor metabolism by controlling saccharification reaction rate in simultaneous saccharification and fermentation of Chinese Maotai-flavor liquor. Int. J. Food Microbiol. **200**, 39–46 (2015)
6. Zhu, Z., Huang, Y.: Population structure diversity of culturable mold in main brewing areas of Maotai Maotai Maotai Baijiu in different rounds. Food Sci. **41**, 184–192 (2020)
7. Tian, D., Yan, Z., Wei, J., Guan, J., Zhao, G., Liu, J.: Research progress on microorganisms and flavor substances in the brewing of light flavor Baijiu. China Brewing **40**, 20–25 (2021)
8. Park, E., Chun, J., Cha, C., Park, W.S., Jeon, C.O., Bae, J.W.: Bacterial community analysis during fermentation of ten representative kinds of kimchi with barcoded pyrosequencing. Food Microbiol. **30**, 197–204 (2012)
9. Wang, X., Du, H., Zhang, Y., Xu, Y.: Environmental microbiota drives microbial succession and metabolic profiles during chinese liquor fermentation. Appl. Environ. Microbiol. **84**, 1128–1145 (2018)

10. Wolfe, B.E., Button, J.E., Santarelli, M., Dutton, R.J.: Cheese rind communities provide tractable systems for in situ and in vitro studies of microbial diversity. Cell **158**, 422–433 (2014)

11. Bokulich, N.A., Ohta, M., Lee, M., Mills, D.A.: Indigenous bacteria and fungi drive traditional kimoto sake fermentations. Appl. Environ. Microbiol. **80**, 5522–5531 (2014)

12. Celano, G., De Angelis, M., Minervini, F., Gobbetti, M.: Different flour microbial communities drive to sourdoughs characterized by diverse bacterial strains and free amino acid profiles. Front. Microbiol. **7**, 1770–1786 (2016)

13. Bourrie, B.C.T., Willing, B.P., Cotter, P.D.: The microbiota and health promoting characteristics of the fermented beverage kefir. Front. Microbiol. **7**, 647–655 (2016)

14. Trček, J., Mahnič, A., Rupnik, M.: Diversity of the microbiota involved in wine and organic apple cider submerged vinegar production as revealed by DHPLC analysis and next-generation sequencing. Int. J. Food Microbiol. **223**, 57–62 (2016)

15. Lei, Z.: Analysis of brewing microorganisms in fragrant baijiu by high throughput sequencing technology. Food Ferment. Ind. **41**, 164–167 (2015)

16. Li, X., Ma, E., Yan, L., Meng, H., Du, X.W., Zhang, S., et al.: Bacterial and fungal diversity in the traditional Chinese liquor fermentation process. Int. J. Food Microbiol. **146**, 31–37 (2011)

17. Wang, H., Huang, Y., Huang, Y.: Microbiome diversity and evolution in stacking fermentation during different rounds of Jiang-flavoured Baijiu brewing. Food Sci. Technol. **143**, 111–119 (2021)

18. Chen, X., Zhang, Y., Yan, Z., Meng, Q., Qi, H., Li, H., et al.: Succession of microbial community in fermented grains of fengxiang type and its correlation with physical and chemical indexes. Food Sci. **41**, 200–205 (2020)

19. Dai, Y., Li, Z., Tian, Z.: Analysis of fungal diversity in daqu and fermented grains of maotai flavor baijiu. Mod. Food Sci. Technol. **34**, 97–104 (2018)

20. Wang, Z., Lu, Z., Shi, J., Xu, Z.: Exploring flavour-producing core microbiota in multispecies solid-state fermentation of traditional Chinese vinegar. Sci. Rep. **6**, 268–286 (2016)

21. Wang, X., Xu, Y., Fan, W., Zhong, F., Tang, Y., Hu, J., et al.: Study on volatile aroma components of luzhou flavor traditional liquor. Brewing Technol. **32**, 31–38 (2013)

22. Liu, J., Zhang, X., Shen, J.: Integrative analysis of the transcriptome and metabolome reveals the mechanism of saline-alkali stress tolerance in Astragalus membranaceus (Fisch) Bge. var. mongholicus (Bge.) Hsiao. Food Qual. Saf. **6**, 1–13 (2022)

23. Han, H., Jiang, X.: Disease biomarker query from RNA-seq data. Cancer Inform. **13**, CIN-S13876 (2014)

EnhanciGraph: Visualizing Enhancer-Gene Interactions

Sri Manjusha Tella$^{(\boxtimes)}$ and Mary Lauren Benton

Baylor University, Waco, TX 76798, USA
`srimanjusha_tella1@baylor.edu`

Abstract. The process of mapping regulatory elements, such as enhancers, to their target genes is a challenging, yet crucial goal in modern biology. We have created an interactive graph tool to visualize the landscape of enhancer-gene links across the human genome by leveraging established benchmarks for enhancer-gene connections, including the Benchmark of Candidate Enhancer-Gene Interactions (BENGI) and the Activity by Contact Model (ABC). In this work, we developed a graph visualization tool, EnhanciGraph, to present the relationships between enhancers and target genes. Using EnhanciGraph, we examined the structure and statistics of the generated graphs and analyzed common edges between target genes and enhancers across different human tissues and cell types. Our tool provides researchers with a user-friendly way to understand the complex relationships between enhancers and genes across multiple tissues and human cell types.

Keywords: enhancers · graph visualization · gene regulation

1 Introduction

The appropriate regulation of gene expression is vital across all biological processes and relies heavily on enhancer activity [16]. Enhancers are short sequences of DNA that can activate gene expression irrespective of their orientation and distance from the target gene [15]. Enhancers modulate transcription through interactions with specific transcription factors and co-factors, often mediated by 3D chromatin interactions, exerting precise and context-dependent control of gene expression.

One key objective in regulatory genomics is to map the association between enhancers and genes to better understand the mechanisms underlying the regulation of gene expression [9]. Greater insight into the regulatory effects of specific enhancer-gene associations could improve our understanding of the regulatory capability of specific enhancers and improve models of gene regulation. Identifying common enhancers and enhancer-gene links across diverse cell types and tissues is also crucial as enhancers may be active across cell types [11]. However, accurately annotating enhancers with their target genes is a difficult task due to the vast search space of the genome, the potential many-to-many relationships between enhancers and genes, and the dynamic, cell-type specific activity of enhancers [1,6,16].

H. Han (Ed.): SDSC 2024, CCIS 2158, pp. 120–134, 2024.
https://doi.org/10.1007/978-3-031-67871-4_9

Current models of enhancer function suggest that three-dimensional chromatin loops bring the enhancer sequences in close physical proximity to the promoter of the target gene, allowing them to regulate gene expression [10,18]. This understanding of enhancer activity aids in comprehension of the role these regulatory elements play in human phenotype. For example, Borsari et al. [2] highlight the impact of enhancers in regulating tissue-specific gene expression and maintaining tissue function. The study shows that enhancers performing tissue-specific functions are mostly located in the intronic regions, while the enhancers associated with housekeeping genes are more often found in intergenic regions. Additionally, they find a shift in the location of enhancers from intergenic to intronic regions as tissues progress through different developmental stages. Research conducted by Nasser et al. [14] revealed insights into the molecular mechanisms of various diseases and complex traits by linking thousands of GWAS signals to specific genes through enhancer-gene connections, emphasizing the link between gene regulation and disease mechanisms.

Several tools currently exist to integrate gene annotations with gene regulatory features. The GenomicKB platform developed by Fan et al. [3] aims to integrate comprehensive knowledge of the human genome, epigenome, transcriptome, and 4D nucleome into a cohesive knowledge graph. This knowledge graph emphasizes relationships and explicit connections between entities. It has a flexible schema and uses ontologies that make it easy to update with new information. Additionally, it integrates external identifiers, which enhances interoperability with other biomedical knowledge graphs. Another web interface tool, PEREGRINE, was introduced by Caitlin et al. [12]. PEREGRINE provides experiemental evidence for enhancer-gene links and allows users to query by target gene or enhancer location.

Cytoscape, as introduced by Shannon et al. [17], serves as a versatile open-source software tailored for generating biological interaction networks and integrating them with gene expression data and other molecular features. Extensions for Cytoscape have been built to help researchers visualize and interpret these networks. For example, work conducted by Janky et al. [8] developed iRegulon, a Cytoscape plugin that deciphers transcriptional regulatory networks and identifies their master regulators and downstream targets, from co-expressed gene sets. Heer et al. [7] developed a novel method called Esearch3D which uses network theory to identify active enhancers based on the dynamic interactions within chromatin architecture.

However, many of these gene regulatory network extensions are focused on protein interactions, rather than DNA sequence interactions. Furthermore, most existing tools do not provide an intuitive way to summarize enhancer-gene interactions within and across tissues. To help fill this gap, we developed an interactive visualization tool, EnhanciGraph. This tool leverages previously annotated enhancer-gene relationships in two curated datasets and provides users with interactive graphs, allowing them to explore the intricate gene regulatory network of a tissue or cell type. Our tool adds additional value by providing intuitive graphs to visualize the enhancer-gene interactions.

These graphs can also be analyzed to derive insights into the properties of the connections between regulatory elements and their target genes. Researchers can use the knowledge gained from our tool about the genomic landscape to answer meaningful biological questions. For example, they can investigate changes in enhancer-gene targeting across tissues, explore regulatory mechanisms during cellular development, study how gene expression contributes to the onset and progression of diseases, and identify potential targets for therapeutic interventions.

2   Methods

2.1   BENGI Dataset

The Benchmark of Candidate Enhancer-Gene Interactions (BENGI) dataset [13] is a curated dataset available for the analysis of regulatory relationships. BENGI derives the enhancer-gene links by integrating the Encyclopedia of DNA Elements (ENCODE) candidate cis-regulatory elements (cCREs) with experimentally derived genomic interactions, providing a benchmark for computational methods [13,19]. Their final database is comprised of the union of predictions derived from the overlap between enhancers and expression quantitative trait loci (eQTL), 3D genomic interactions, and CRISPR experiments. We downloaded the All-Pairs. Natural-Ratio data files from the BENGI GitHub repository for all available tissues. The data initially consisted of cCRE accession numbers and gene identifiers. For interoperability, we converted the cCRE accession numbers to genomic coordinates using the GRCH37/hg19 assembly of the human genome. By using tissue-specific information available in BENGI, we partitioned the enhancer-gene links based on both genomic location and active tissue, allowing for analysis of regulatory relationships across different chromosomes within specific tissues.

2.2   ABC Dataset

The Activity-by-Contact (ABC) model [5] predicts enhancer-gene links by estimating enhancer activity and contact frequency with promoters using epigenomic data. The ABC model uses chromatin state measurements to create detailed maps of enhancer-gene connections in specific cell types. We downloaded the data from both the ABC-Max [14]. The ABC-Max data contained 131 unique cell types and 513 unique target genes. In addition to ABC-Max data, we also integrated the ABC Prediction dataset to generate precomputed graphs with all enhancer-gene connections across 131 cell types [5]. In addition to the genomic coordinates of the enhancer locations, we also consider the ABC score associated with each enhancer-gene link. The graphs constructed for the ABC model represent the enhancers and target genes as nodes with edges weighted by the ABC score.

To enhance the interpretability of the graphs, we color-coded the nodes and varied the edge thickness based on three categories of the ABC score: low, medium, and high. We categorized the numerical ABC score into specified groups based on their percentiles (divided into thirds) to enable easy interpretation and analysis. The enhancer nodes are in light blue, while the target genes are in green. ABC scores are indicated with three colors: red for low, yellow for medium, and green for high. This distinction allows for the quick visual identification of the potential components of interest within the regulatory networks.

2.3   EnhanciGraph Architecture

The EnhanciGraph tool is a web-based application designed to explore the significance of enhancer-gene interactions across diverse cell types. The architecture of the tool encompasses the steps of data preprocessing, graph construction, and interactive visualization. We preprocess the data by first integrating the enhancer and gene annotations from two benchmark datasets: BENGI and ABC. Using PyVis, we construct graphs by building adjacency matrices for each cell type based on the existing enhancer-gene links. The tool stores precomputed graphs for both BENGI tissue-specific and ABC cell-specific enhancer gene links locally. The front end of the application was built using HTML, CSS, and JavaScript, allowing for a user-friendly interface for selecting cell types and exploring overlapping links. When users make selections in the overlapping edge explorer component, the back end processes the data to identify overlapping links using the Flask framework. It then generates and renders the graph visualization in association with user selection and graph legend. Additionally, the architecture includes features for scalability and performance optimization, ensuring smooth functionality even with large datasets. Overall, our platform provides a seamless experience for researchers to explore and analyze enhancer-gene interactions across different cell types and tissues. The code can be accessed on GitHub: https://github.com/sri-tella/EnhanciGraph.

2.4   System Requirements

Our tool's graph visualizations are developed using Python libraries and functions, including NetworkX (version 2.8.4) for graph creation and manipulation, and pyVis (version 0.3.2) for rendering interactive graphs in an HTML format. We used two benchmark datasets: ABC, which is approximately 1.84 GB, and BENGI, which is 83.3 MB in size. The graph construction and rendering were performed on a system with 8 GB RAM and macOS. Although it was tested on macOS, our tools could be deployed across platforms.

2.5   Analysis of Graphs

We compute several graph statistics to provide quantitative insights into the network properties of the gene regulatory networks. We include common statistics such as the number of nodes and edges, the density of the graph, and the assortativity. The number of nodes and edges refers to the total number of enhancers, genes, and their interactions in a specific graph. The density of the graph is a metric that quantifies the ratio of the existing edges to the total possible number of edges. A higher density indicates a more interconnected graph. The assortativity measures the tendency of nodes to connect to other nodes with similar attributes. It assesses whether nodes with similar characteristics are more likely to be connected than nodes with different characteristics.

3   Results

3.1   BENGI Data

In our analysis of the BENGI dataset, we observed a higher number of genes relative to enhancers in the constructed graphs (Fig. 1). We observed significant differences between BENGI tissues, ABC, and ABC-Max across all cell types. Table 1 provides a summary of our findings, which include variation in the number of enhancers and genes, as well as the average number and proportion of linked genes and enhancers per tissue.

Table 1. Summary of BENGI, ABC, and ABC-Max data

Metric	BENGI	ABC	ABC-Max
Unique number of tissues	6	131	131
Unique number of enhancers	9,484	3,127,243	101
Unique number of genes	41,911	23,227	513
Average number of genes per tissue	32.15	14,325.42	27.08
Average proportion of linked genes per tissue	0.002	2.53	1.00
Average number of enhancers per tissue	2.338	25,162.20	9.22
Average proportion of linked enhancers per tissue	0.004	0.018	1.00

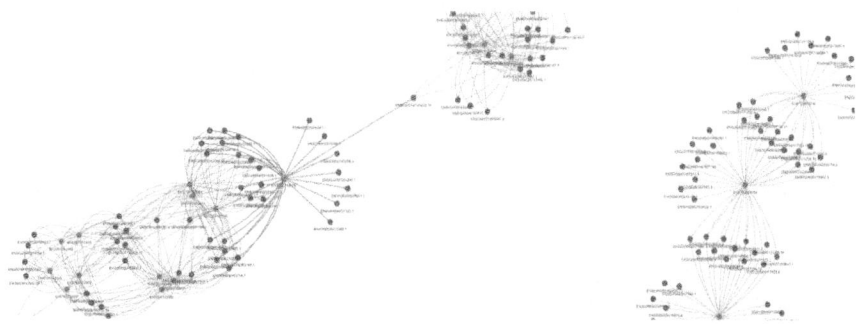

Fig. 1. Screenshot of a graph of chromosome 1 for liver tissue in BENGI dataset. Red nodes represent enhancers, while blue nodes represent genes. (Color figure online)

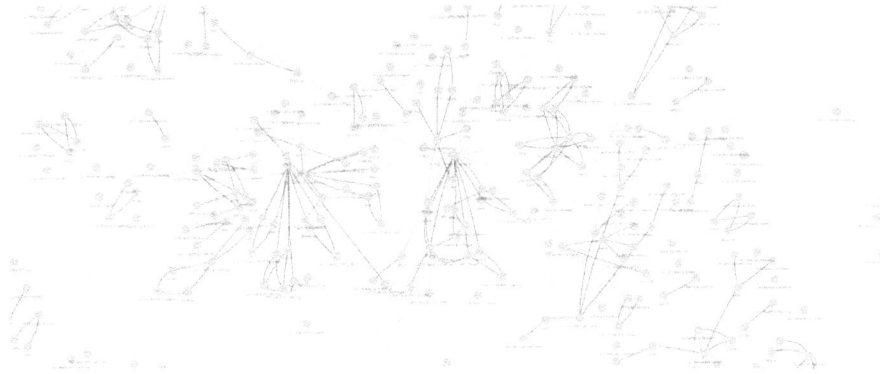

Fig. 2. Screenshot of a graph of chromosome 1 for liver ENCODE tissue in ABC predictions dataset. Blue nodes represent enhancers, while green nodes represent genes. The edges of the graph are color-coded based on their weight, with high scores in green, medium in yellow, and low in red. (Color figure online)

3.2   ABC Data Analysis

The graphs generated from the full ABC dataset have more enhancers than genes, likely due to the more expansive definition for the enhancer sequences and a less conservative method for enhancer-gene links (Fig. 2). On the other hand, for the reduced ABC-Max dataset, we found that there were more genes than enhancers, which is consistent with what we observed in BENGI (Fig. 3). To evaluate the similarities and differences between regulatory networks for genes relevant to disease, we generated a heatmap from the adjacency matrix derived from the ABC-Max dataset. Clustering using both rows and columns revealed common genes across the 131 cell types (Fig. 4). We extracted two clusters comprising 3 and 8 cell types associated with target genes *FUBP1* and *BCL9L*, respectively. Both *FUBP1* and *BCL9L* are potentially important genes in the study of cancer. Altered expression of *FUBP1* has been found in tumor cells, while *BCL9L* is involved in TGF-β signaling and could play a role in cancer metastasis. To better understand the cross-tissue regulatory control of these genes, we constructed two combined graphs for each cluster constituting all enhancer-gene links to identify overlapping links (depicted by red colored edges in the graph) (Fig. 5). We found 13 common enhancer-gene associations across 3 cell types for the *FUBP1* cluster, and 26 common enhancer-gene associations across the 8 cell types in the *BCL9L* cluster. Our graph highlighted that genes that are active across multiple tissues can have both shared and unique regulatory elements. The shared elements suggest some conserved regulatory structure across tissues with similar expression patterns, while the unique elements could provide tissue-specific control of expression level. This could be particularly relevant for the genes, such as *FUBP1* and *BCL9L*, in the ABC-Max dataset, where these regulatory networks could improve our understanding of genes involved in disease processes.

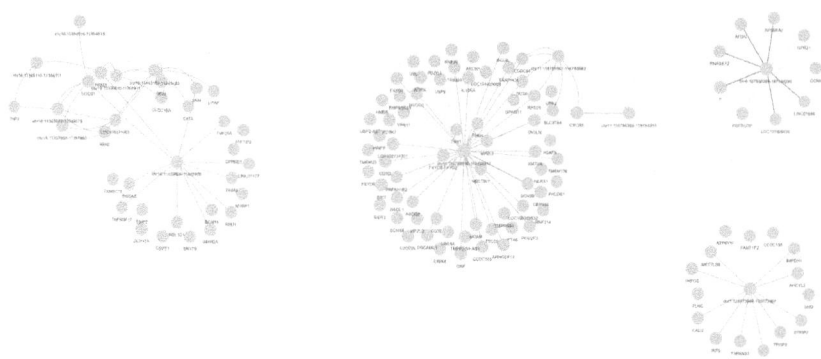

Fig. 3. ABC-Max dataset graph. Light blue nodes represent enhancers, while green nodes represent genes. (Color figure online)

3.3   Case Studies

Below we present a series of examples to demonstrate two potential use cases for our tool.

Example Usage 1: Tissue/Cell-Specific Analysis. Our tool facilitates tissue-specific analysis by allowing researchers to explore genomic data from the BENGI dataset. Users can browse tissue-specific files and visualize the interactive graphs at the chromosome level to uncover enhancer-gene links in each tissue, as seen in Fig. 6. The rendered graph visualizations consist of enhancers and target genes represented by red and blue nodes, respectively. This functionality enables researchers to gain insights into regulatory networks across tissues.

Researchers can select specific cell types of interest and browse the interactive graph visualizations of the associated regulatory network. For example, when we analyze the liver tissue from ENCODE [19] in the ABC-Max dataset, our graph displays 3 enhancers and 28 genes. Many of the genes have few connected enhancers, suggesting that there is simpler regulatory control of expression for these genes in the liver. In contrast, for the pancreas tissue from ENCODE [19], we observe 11 enhancers and 14 genes. Most of these genes have many enhancers. This feature enables researchers to delve deeper into cell-specific regulatory mechanisms and identify key genes or relevant cellular pathways (Fig. 7). The visual representations of these graphs depict the enhancer location and tar-

Fig. 4. ABC-Max dataset cluster map with unique target genes (513) on the x-axis and unique cell types (131) on the y-axis. The color gradient indicates the number of edges associated with a specific cell type and target gene.

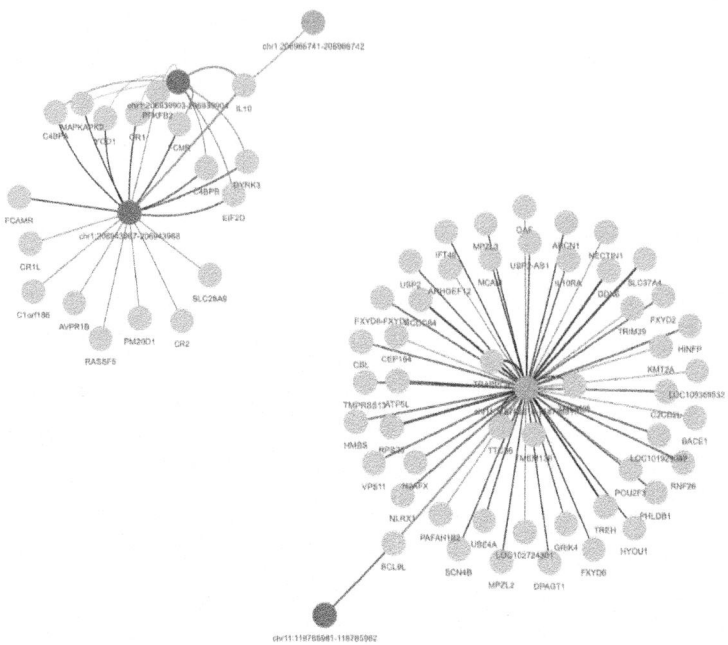

Fig. 5. Graph representing the overlapped edges of an extracted cluster with 8 cell types associated with *BCL9L*: GM12878, two dendritic cells (treated for 4 or 6 h), CD19 B cells, CD14 monocytes, breast epithelium, H9, and thymus. Colored nodes represent enhancers in each cell type, while grey nodes represent genes and red color edges represent the overlapping edges among the cell types. (Color figure online)

get genes as nodes colored in light blue and green, respectively. The edges are weighted based on the ABC score, which is categorized into high (green), medium (yellow), and low (red).

Example Usage 2: Edge Overlap Analysis. One of the unique capabilities of our tool is edge overlap analysis, which allows researchers to investigate overlapping regulatory interactions between different cell types in the ABC-Max dataset. Researchers can better understand shared regulatory mechanisms by comparing overlapped cell-specific graph visualizations and identifying common

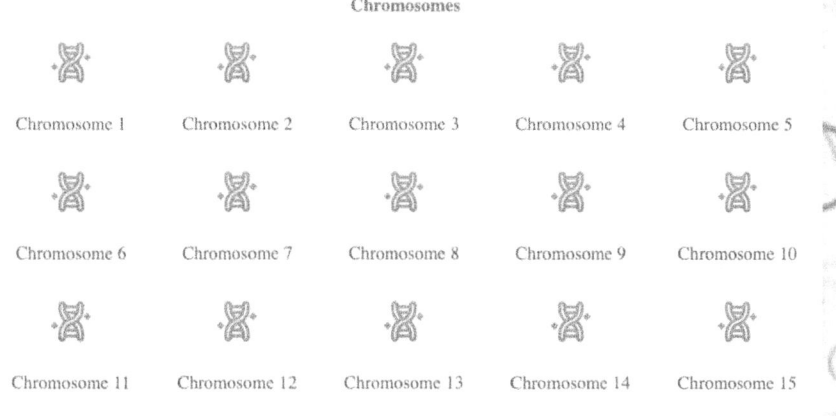

Fig. 6. EnhanciGraph user interface for liver tissue. Clicking on a chromosome displays a graph of genes and enhancers specific to the selected liver tissue chromosome.

regulatory elements across cell types. With this feature, users can select up to five cell types and explore the overlapping links through an interactive graph (Fig. 8). This graph consolidates data from the chosen cell types and presents a comprehensive view of enhancer gene interactions. Enhancers annotated each cell type in the newly constructed graph are colored by cell type, while grey nodes represent the genes. The edges highlighted in red within the graph make it easier to identify overlapping links across the selected cell types, allowing users to identify shared regulatory interactions.

1. **Data Curated from Git Repo:** The ABC dataset is sourced from a Git repository, ensuring up-to-date and relevant data for analysis.
2. **Graph Generation:** Graphs are generated by considering chromosome start and end as enhancer regions, with associated target genes forming nodes.
3. **ABC Score-Based Edges:** Graph edges between enhancers and genes are determined by ABC scores.
4. **Color Mappings:** Nodes are color-coded, with light blue representing enhancer regions and green representing target genes.

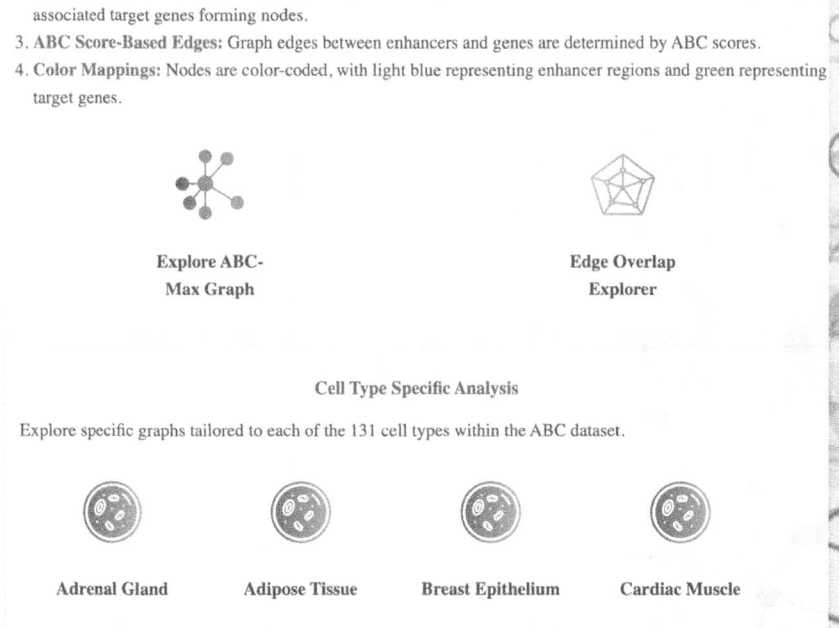

Explore ABC-
Max Graph

Edge Overlap
Explorer

Cell Type Specific Analysis

Explore specific graphs tailored to each of the 131 cell types within the ABC dataset.

Adrenal Gland   **Adipose Tissue**   **Breast Epithelium**   **Cardiac Muscle**

Fig. 7. EnhanciGraph user interface to explore cell-specific graphs. Choosing a specific cell type displays the corresponding graph. The "Explore ABC-Max" graph component shows all enhancer and gene interactions between all cell types. The "Edge Overlap Explorer" component helps to highlight overlapping interactions between selected cell types.

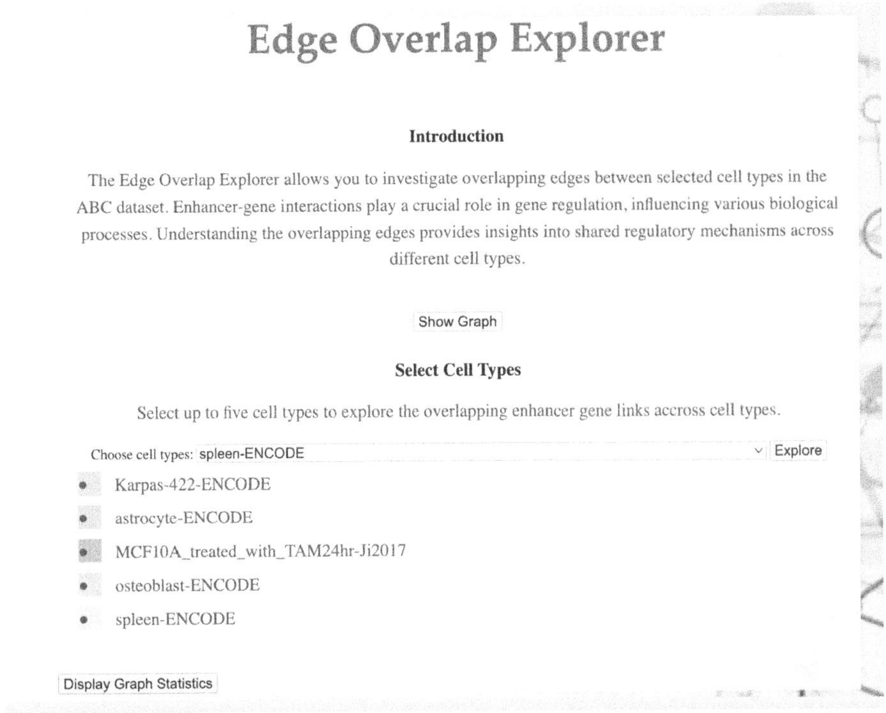

Fig. 8. User interface for the edge overlap explorer. The user can select up to five cell types from a drop-down menu. The graph generated by clicking the "Explore" button combines enhancer-gene links data across all selected cell types. Cell types are color-coded to represent specific enhancers. The "Show Graph" component renders the constructed graph. This graph highlights the overlapped edges in red as shown in Fig. 5. The "Display Graph Statistics" component showcases the number of enhancers, target genes, edges, and common genes across cell types. (Color figure online)

4 Discussion

We developed a web tool, EnhanciGraph, to visualize enhancer-gene networks across diverse human tissues and cell types. In addition to interactive visualizations, EnhanciGraph also offers the Edge Overlap Explorer, which enables users to observe overlapping enhancer-gene interactions across selected cell types. This tool can provide insights into common regulatory relationships across diverse cellular contexts. Finally, EnhanciGraph's user interface and analysis tools are simple and require no programming, increasing accessibility for a broad scientific audience.

Our tool utilizes existing enhancer-gene maps; as such, it is constrained by the limitations of these datasets. Although the BENGI dataset is the first benchmark dataset of its kind [13], it may still contain false positives and could be missing many real enhancer-gene links due to the data used during construction.

Similarly, the ABC model was developed using a CRISPR-based approach to identify true enhancer-gene targets but may not capture all relevant regulatory interactions in a given cell type. Our results highlight the importance of continued improvement in computational methods for enhancer-gene prediction, and BENGI serves as a valuable framework for evaluating these enhancer-gene links across the human genome. In the future, we plan to expand our tool by incorporating data from databases of predicted enhancer-gene links, such as GeneHancer [4]. Furthermore, we envision expanding EnhanciGraph's functionality by enabling researchers to upload their own enhancer-gene link data, thus offering a customizable platform to explore the regulatory landscape of the human genome.

One additional limitation of our approach is the technical challenge associated with generating and rendering large graphs, particularly as datasets grow in size and complexity. As new datasets are developed with increased data volume, the scalability of our graph visualization tool may become a significant concern. Addressing this challenge will require ongoing optimization efforts and exploring other techniques for handling and visualizing massive graph structures.

5   Conclusion

In our study, we leveraged data from two benchmark datasets of experimentally supported enhancer-gene links [5, 13] to develop EnhanciGraph. EnhanciGraph is a user-friendly web-based application designed to provide users with interactive visualizations of gene regulatory networks.

Our tool makes several important contributions. First, our graph visualizations portray the enhancer to gene links in an interactive manner. Using BENGI, we developed tissue-specific graph visualizations for each chromosome, and using the ABC model data we improved those graphs by weighting with the ABC scores. Second, we developed a method to analyze the overlapping links among the graph structures across diverse cell types using the ABC data. This can provide insights into shared regulatory landscapes across tissues. Finally, we provide the Edge Overlap Explorer to visually highlight the overlapping links among selected cell types. By leveraging the power of web-based applications, EnhanciGraph can visualize and interact with enhancer-gene links via graphs, making it accessible and easy to use for researchers in the field of genomics.

In our future work, we plan to improve EnhanciGraph by incorporating additional visualization features and expanding into additional enhancer-gene datasets, such as Genehancer [4]. Although there are existing tools such as Cytoscape [17] available to explore biological networks, EnhanciGraph offers a unique perspective on enhancer-gene links, providing insights that are complementary to such tools by focusing on cross-tissue comparisons. To overcome limitations posed by individual datasets, we plan to ensure regular maintenance and monitoring of our website, ensuring the information provided remains accurate. Furthermore, we recognize the need to address the complexity of the visualization algorithm used. We will continue to explore further optimization strategies to enhance the efficiency and scalability of our tool, thereby improving the usability of EnhanciGraph.

References

1. Andersson, R., Gebhard, C., Miguel-Escalada, I., et al.: An atlas of active enhancers across human cell types and tissues. Nature **507**(7493), 455–461 (2014). https://doi.org/10.1038/nature12787
2. Borsari, B., Villegas-Mirón, P., Pérez-Lluch, S., et al.: Enhancers with tissue-specific activity are enriched in intronic regions. Genome Res. **31**(8), 1325–1336 (2021). https://doi.org/10.1101/gr.270371.120
3. Feng, F., Tang, F., Gao, Y., et al.: Genomickb: a knowledge graph for the human genome. Nucleic Acids Res. **51**(D1), D950–D956 (2023). https://doi.org/10.1093/nar/gkac957
4. Fishilevich, S., Nudel, R., Rappaport, N., et al.: Genehancer: genome-wide integration of enhancers and target genes in genecards. Database (Oxford) **2017**, bax028 (2017). https://doi.org/10.1093/database/bax028
5. Fulco, C.P., Nasser, J., Jones, T.R., et al.: Activity-by-contact model of enhancer-promoter regulation from thousands of crispr perturbations. Nat. Genet. **51**, 1664–1669 (2019). https://doi.org/10.1038/s41588-019-0538-0
6. Gasperini, M., Tome, J.M., Shendure, J.: Towards a comprehensive catalogue of validated and target-linked human enhancers. Nat. Rev. Genet. **21**, 292–310 (2020). https://doi.org/10.1038/s41576-019-0209-0
7. Heer, M., Giudice, L., Mengoni, C., Giugno, R., Rico, D.: Esearch3d: propagating gene expression in chromatin networks to illuminate active enhancers. Nucleic Acids Res. **51**(10), e55 (2023). https://doi.org/10.1093/nar/gkad229
8. Janky, R., et al.: iregulon: from a gene list to a gene regulatory network using large motif and track collections. PLoS Comput. Biol. **10**(7), e1003731 (2014)
9. Kim, D., An, H., Shearer, R.S., et al.: A principled strategy for mapping enhancers to genes. Sci. Rep. **9**(1), 11043 (2019). https://doi.org/10.1038/s41598-019-47521-w
10. Krivega, I., Dean, A.: Enhancer and promoter interactions-long distance calls. Curr. Opin. Genet. Dev. **22**(2), 79–85 (2012)
11. Laiker, I., Frankel, N.: Pleiotropic enhancers are ubiquitous regulatory elements in the human genome. Genome Biol. Evol. **14**(6), evac071 (2022). https://doi.org/10.1093/gbe/evac071
12. Mills, C., et al.: Peregrine: a genome-wide prediction of enhancer to gene relationships supported by experimental evidence. PLoS ONE **15**(12), e0243791 (2020)
13. Moore, J.E., Pratt, H.E., Purcaro, M.J., et al.: A curated benchmark of enhancer-gene interactions for evaluating enhancer-target gene prediction methods. Genome Biol. **21**(1), 17 (2020). https://doi.org/10.1186/s13059-019-1924-8
14. Nasser, J., Bergman, D.T., Fulco, C.P., et al.: Genome-wide enhancer maps link risk variants to disease genes. Nature **593**, 238–243 (2021). https://doi.org/10.1038/s41586-021-03446-x
15. Panigrahi, A., O'Malley, B.: Mechanisms of enhancer action: the known and the unknown. Genome Biol. **22** (2021). https://doi.org/10.1186/s13059-021-02322-1
16. Pennacchio, L.A., Bickmore, W., Dean, A., Nobrega, M.A., Bejerano, G.: Enhancers: five essential questions. Nat. Rev. Genet. **14**(4), 288–295 (2013). https://doi.org/10.1038/nrg3458

17. Shannon, P., et al.: Cytoscape: a software environment for integrated models of biomolecular interaction networks. Genome Res. **13**(11), 2498–2504 (2003)
18. Shlyueva, D., Stampfel, G., Stark, A.: Transcriptional enhancers: from properties to genome-wide predictions. Nat. Rev. Genet. **15**(4), 272–286 (2014). https://doi.org/10.1038/nrg3682
19. The ENCODE project consortium: an integrated encyclopedia of DNA elements in the human genome. Nature **489**, 57–74 (2012). https://doi.org/10.1038/nature11247

Event-Triggered Control for Synchronization of Chaotic Delayed Lur'e Systems with Stochastic Cyber-Attacks

Yanjie Liu and Zhi Li[✉]

School of Mechano-Electrionic Engineering, Xidian University, X'ian 710071, China
`yanjieliu@stu.xidian.edu.cn, zhli@xidian.edu.cn`

Abstract. This paper studies the problem of event-triggered controller design for chaotic delayed Lur'e systems with stochastic cyber-attacks. Firstly, a synchronization error model is proposed to present the synchronization of the master-slave system. Secondly, in order to save the energy consumption and reduce the transmission load of network, the event-triggered scheme is constructed. Thirdly, on the basis of Lyapunov stability theory, sufficient conditions for stabilization which can ensure synchronization of the master system and slave system with stochastic cyber-attacks are obtained, and the precise expression of the controller gain is given. Finally, an example is presented to illustrate the effectiveness of the proposed method.

Keywords: chaotic delayed Lur'e system · event-triggered · synchronization · stochastic cyber-attacks

1 Introduction

In the field of control, chaotic Lur'e systems are frequently used to delegate a class of linear dynamic elements and nonlinear feedback systems based on sector constraints. The well-known chaotic Lur'e system includes Chua's circuit, chaotic attractors and neural networks [1–3]. Since Carroll and Pecora conducted pioneering research on the synchronization of chaotic systems in 1991 [4], the synchronization of chaotic systems has attracted attention of many scholars and has been applied in many fields such as secure communication, lasers and life sciences. For delayed chaotic Lur'e systems, whose purpose is to make the output of the slave system follow the output of the master system [5].

In recent years, with the development of networks and modern technology, network control systems exchange information through a shared communication network. In the process of data transmission, the existence of factors such as limited network bandwidth and cyber-attacks will not only reduce the transmission capacity of the communication network, but also deteriorate the performance of the network system [6]. Time-triggered and event-triggered control algorithms are two more effective methods to reduce the burden of network transmission

H. Han (Ed.): SDSC 2024, CCIS 2158, pp. 135–147, 2024.
https://doi.org/10.1007/978-3-031-67871-4_10

[7]. However, the time-triggered control method has an insurmountable short-coming, that is, a fixed sampling interval will be selected in the system, the controllers of all agents need to be updated synchronously, which will waste computing and battery equipment energy resources. The event-triggered scheme can effectively reduce the communication load of the network while maintaining the system performance [8]. Literature [9] designed hybrid-triggered controller with quantization to reduce the data transfer rate of the Takagi-Sugeno (T-S) fuzzy system under cyber-attacks. Literature [10] studied the synchronization problem of event-triggered delayed neural networks. Literature [11] employed the sampled data control method to deal with the asymptotical synchronization problem of two identical delayed chaotic Lur'e system. Literature [12] studied a sampling data synchronization method of delayed Lur'e system under distur-bance. In order to reduce the communication burden, literature [13] introduced an aperiodic event-triggered transmission scheme to determine the transmis-sion of the latest sampled synchronization data, and studied the master-slave synchronization of the chaotic Lur'e system under aperiodic sampled data. A novel memory-based event-triggered mechanism with the past information of the system is presented for networked chaotic Lur'e systems in [14]. [15] inves-tigated the global synchronization issue of the chaotic delayed Lur'e system via memory-based event-triggered control. With the increasing integration of net-worked communication elements and physical components, the systems become increasingly vulnerable to cyber-attacks. Essentially, cyber-attacks are methods or techniques that target physical processes through the network. By manipu-lating sensory data or controlling actuator inputs, adversaries can disrupt sys-tem performance or even redirect the plant as desired [16]. Unfortunately, the problem of event-triggered control for synchronization of chaotic delayed Lur'e systems with stochastic cyber-attacks has not been solved yet, which inspired our current research.

In this paper, we consider the event-triggered control problem of the master-slave synchronization of the chaotic time-delay Lur'e system attacked by the net-work. In order to save network resources, an event-triggered scheme is adopted to judge whether the current sampled data is transmitted through the network. Meanwhile, a synchronization error model is proposed for the discussed system based on the above discrete-time event-triggered scheme. Lyapunov stability theory is used to derive the sufficient conditions for the stability of the synchro-nization error system, and the required controller design method is given based on the solution of linear matrix inequalities (LMIs). Finally, an example is used to verify the effectiveness of the designed method.

2   Preliminaries and Problem Statement

2.1   Preliminaries

$\mathbb{R}^n$ and $\mathbb{R}^{n \times m}$ denote the n-dimensional real vector space and $n \times m$ real matrix space, respectively. I means the identity matrix. $\mathbf{0}_{n \times m}$ denotes an $n \times m$ zero matrix. $diag\{U_1, ..., U_n\}$ defines a block-diagonal matrix with matrices $U_i, i = 1, ..., n$ on its diagonal elements. For a real symmetric matrix U, $U > 0$ ($U < $

0) reprsents a positive definite (negative definite) matrix. U^T and U^{-1} denote transposition and inverse of matrix U, respectively. $*$ in symmetric block matrices is used to represent the symmetric elements.

2.2 The Master-Slave Chaotic Delayed Lur'e System

Consider the following general master-slave type of chaotic delayed Lur'e system with cyber-attacks

$$\begin{cases} M : \dot{m}(t) = Am(t) + Bm(t - \tau_1(t)) + W\sigma(Dm(t)), \\ S : \dot{s}(t) = As(t) + Bs(t - \tau_1(t)) + W\sigma(Ds(t)) + u(t), \end{cases} \tag{1}$$

where M, S represent the master system and the slave system, respectively. $m(t), s(t) \in \mathbb{R}^n$ are the state vectors of two systems, $\sigma(\cdot) \in \mathbb{R}^m$ is a nonlinear function vector, $u(t) \in \mathbb{R}^m$ is control input vector. A, B, W are system matrices, $D = [D_1, D_2, \cdots, D_m]^T$ is a row vector. The time delay $\tau_1(t)$ is bounded as $0 < \tau_1(t) < \tau_1$, τ_1 is the upper bound on the delay.

Remark 1. Non-zero lower delay $\tau_1(t)$ and nonlinear function $\sigma(\cdot)$ are used due to the system expressed by the above formula is more realistic and can be widely used.

Assumption 1. *It is assumed that function $\sigma_i(\cdot)$ in the system (1) is a continuous bounded diagonal nonlinear function, $\forall q_1, q_2 \in \mathbb{R}, q_1 \neq q_2$, satisfies*

$$l_i^- \leq \frac{\sigma_i(q_1) - \sigma_i(q_2)}{q_1 - q_2} \leq l_i^+, i = 1, 2, \cdots, m, \tag{2}$$

where l_i^-, l_i^+ are all known constants.

2.3 The Synchronization Error System and Event-Triggered Scheme

The synchronization error is defined as $x(t) = s(t) - m(t)$. The synchronization error system is given by

$$\dot{x}(t) = Ax(t) + Bx(t - \tau_1(t)) + W\sigma(Dx(t)) + u(t). \tag{3}$$

According to Assumption 1, the nonlinear function $\phi_i(D_i x(t)) = \sigma_i(D_i s(t)) - \sigma_i(D_i m(t))$ satisfies

$$l_i^- \leq \frac{\sigma_i(D_i m(t)) - \sigma_i(D_i s(t))}{D_i m(t) - D_i s(t)} = \frac{\phi_i(D_i x(t))}{D_i x(t)} \leq l_i^+, i = 1, 2, \cdots, m. \tag{4}$$

The data transmitted on the network is vulnerable to attacks because the hacker's purpose is to attack the controller by modifying the control input. Considering cyber-attacks in the design of controller, the controller can be defined as

$$u(t) = Kx(t_k h) + \beta(t)K[\alpha(t)g(x(t-d(t))) + (1-\alpha(t))h(x(t-\eta(t))) - x(t_k h)], \tag{5}$$

where $d(t) \in [0, d], \eta(t) \in [0, \eta]$ represent time-delay of cyber-attacks, d, η denote the maximum of time-delay, $\alpha(t), \beta(t)$ take values on $\{0, 1\}$.

Remark 2. In this paper, cyber-attacks initiated by hackers are set as a function $\alpha(t)g(x(t - d(t))) + (1 - \alpha(t))h(x(t - \eta(t)))$, which is more consistent with the actual situation. $g(x(t - d(t))$ and $h(x(t - \eta(t)))$ represent the different characteristics of the cyber-attacks. $\alpha(t) = 1$ means the cyber-attacks function is $g(x(t - d(t)))$, $\alpha(t) = 0$ means the cyber-attacks function is $h(x(t - \eta(t)))$. Hackers will modify the transmitted data to achieve cyber-attacks and eventually lead to system instability. Therefore, it is of great significance to design the corresponding control method to defend against the cyber-attacks.

Remark 3. Due to restricted access to system information, hackers may not be able to successfully modify the transmitted data, so the control system may be subject to intermittent cyber-attacks. $\beta(t)$ is used to describe the presence of cyber-attacks. $\beta(t) = 1$ indicates that the cyber-attacks was successfully implemented. $\beta(t) = 0$ represents that the data is transmitted normally.

It is assumed that $\alpha(t)$ and $\beta(t)$ are independent identically distributed random variables, satisfying

$$Pr\{\alpha(t) = 1\} = \bar{\alpha}, Pr\{\beta(t) = 1\} = \bar{\beta}, E\{\alpha\} = \bar{\alpha}, E\{\beta\} = \bar{\beta},$$
$$E\{(\alpha(t) - \bar{\alpha})^2\} = \bar{\alpha}(1 - \bar{\alpha}) = \rho_1^2, E\{(\beta(t) - \bar{\beta})^2\} = \bar{\beta}(1 - \bar{\beta}) = \rho_2^2,$$
$$E\{(\alpha(t) - \bar{\alpha})(\beta(t) - \bar{\beta})\} = 0.$$

In order to reduce the network transmission burden, a distributed event-triggered method is introduced into the synchronization error system.

$$\|\Omega^{\frac{1}{2}}[x(t_k h + lh) - x(t_k h)]\|_2 \leq \epsilon\|\Omega^{\frac{1}{2}}x(t_k h + lh)\|_2, \tag{6}$$

where $\epsilon \in [0, 1]$ is a predefined constant, $\Omega > 0$ is a event-triggered weighting matrix to be determined, lh represents the sampling time, and $t_k h$ represents the most recent event-triggered instants.

The structure of distributed event-triggered networked control systems under cyber-attacks is shown in Fig. 1.

Remark 4. Once the event-triggered condition (6) is violated, the latest sampled data will be transmitted. Using the event-triggered scheme to determine whether the data can be transmitted to the quantizer can reduce the load on the network.

Denote

$$e_k(t) = x(t_k h) - x(t_k h + lh), \quad t \in [t_k, t_{k+1})_{k=0}^{\infty}, \tag{7}$$
$$\tau_2(t) = t - t_k h - lh, l = 0, 1, 2, \cdots, 0 \leq \tau_2(t) \leq \tau_2. \tag{8}$$

From (7) and (8), the event-triggered condition can be rewritten as

$$e_k^T(t)\Omega e_k(t) \leq \epsilon x^T(t - \tau_2(t))\Omega x(t - \tau_2(t)). \tag{9}$$

Based on the definitions of $e_k(t)$ and $\tau_2(t)$, the controller (5) can be rewritten as

$$u(t) = (1 - \beta(t))K[x(t - \tau_2(t)) + e_k(t)] + \beta(t)K[\alpha(t)g(x(t - d(t)))$$
$$+ (1 - \alpha(t))h(x(t - \eta(t)))]. \tag{10}$$

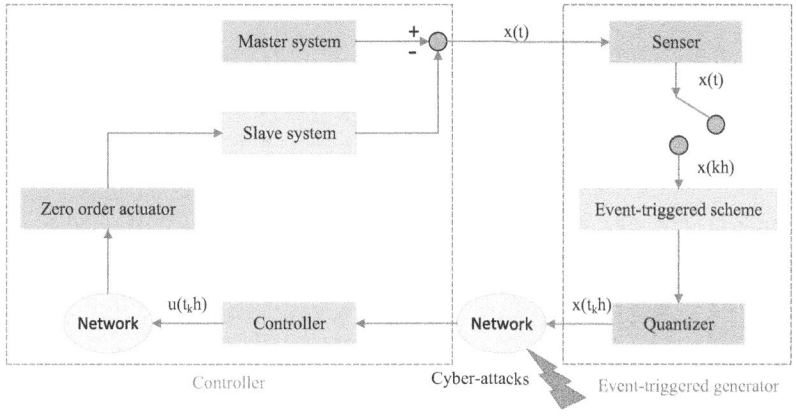

Fig. 1. Distributed event-triggered control systems structure

Substituting (10) into (3) yields

$$\dot{x}(t) = \Pi_0 + (\beta(t) - \bar{\beta})\Pi_1 + \bar{\beta}(\alpha(t) - \bar{\alpha})\Pi_2 + (\beta(t) - \bar{\beta})(\alpha(t) - \bar{\alpha})\Pi_2$$
$$+ (\bar{\beta} - \beta(t))K[x(t - \tau_2(t)) + e_k(t)], \tag{11}$$

where $\Pi_0 = Ax(t) + Bx(t - \tau_1(t)) + W\phi(Dx(t)) + (1 - \bar{\beta})K[x(t - \tau_1(t)) + e_k(t)] + \bar{\beta}\Pi_1, \Pi_1 = \bar{\alpha}Kg(x(t - d(t))) + (1 - \bar{\alpha})Kh(x(t - \eta(t))), \Pi_2 = Kg(x(t - d(t))) - Kh(x(t - \eta(t)))$.

Assumption 2. *The stochastic cyber-attacks $g(x(t))$ and $h(x(t))$ are assumed to be nonlinear functions satisfying*

$$\|g(x(t))\|_2 \leq \|Gx(t)\|_2, \tag{12}$$

$$\|h(x(t))\|_2 \leq \|Hx(t)\|_2, \tag{13}$$

where G and H are known constant matrices representing the upper bounds of the nonlinearities.

Lemma 1. *[17] For any matrices Y and positive definite matrices X satisfying $\Gamma = \begin{bmatrix} X & Y \\ * & X \end{bmatrix} \geq 0$, positive scalars δ_1, δ_2 satisfying $\delta_1 + \delta_2 = 1$, the following inequality holds*

$$-\frac{1}{\delta_1}\omega_1^T X\omega_1 - \frac{1}{\delta_2}\omega_2^T X\omega_2 \leq -\begin{bmatrix} \omega_1 \\ \omega_2 \end{bmatrix}^T \Gamma \begin{bmatrix} \omega_1 \\ \omega_2 \end{bmatrix}. \tag{14}$$

Lemma 2. *[18] For matrices $R > 0$ and $X^T = X$, satisfying*

$$-XR^{-1}X \leq k^2 R - 2kX, \tag{15}$$

where k is an arbitrary constant.

3   Main Results

In this section, sufficient conditions are derived to ensure the stability of event-triggered synchronization error system (3) with stochastic cyber-attacks in Theorem 1. Then, the controller design problem is solved and the controller gain can be computed based on Theorem 1.

Theorem 1. *For given delay parameters $\tau_1 > 0, \tau_2 > 0, d > 0, \eta > 0$ and constants $\bar{\alpha} \in [0,1], \bar{\beta} \in [0,1], \epsilon \in [0,1]$, the event-triggered synchronization error system with stochastic cyber-attacks is asymptotically stable if there exist matrices $P \in \mathbb{R}^{n \times n} > 0, T_i \in \mathbb{R}^{n \times n} > 0, R_i \in \mathbb{R}^{n \times n} (i = 1, \cdots, 4), \Omega \in \mathbb{R}^{n \times n} > 0, \Lambda = diag\{\lambda_1, \lambda_2, \ldots, \lambda_m\}$, scalars $\lambda > 0$ such that*

$$\bar{R}_i = \begin{pmatrix} R_i & T_i \\ * & R_i^T \end{pmatrix} \geq 0, \tag{16}$$

$$\Xi < 0, \tag{17}$$

where
$\Xi = e_1^T P \bar{\Pi}_0 + \bar{\Pi}_0^T P e_1 + e_1^T (Q_1 + Q_2 + Q_3 + Q_4) e_1 - e_2^T Q_1 e_2 - e_4^T Q_2 e_4 - e_5^T Q_3 e_5 - e_3^T Q_4 e_3 + \bar{\Pi}_0^T R \bar{\Pi}_0 + (\bar{\beta}^2 \rho_1^2 + \rho_1^2 \rho_2^2) \bar{\Pi}_2^T R \bar{\Pi}_2 + \rho_2^2 \bar{\Pi}_1^T R \bar{\Pi}_1 - T_1^T \bar{R}_1 T_1 + \rho_2^2 (e_7^T + e_{13}^T) K^T R K (e_7 + e_{13}) - T_2^T \bar{R}_2 T_2 - T_3^T \bar{R}_3 T_3 - T_4^T \bar{R}_4 T_4 - 2 e_{10}^T \Lambda e_{10} - 2 e_1^T D^T \Lambda D e_1 + 2 e_{10}^T (L^+ + L^-) + \bar{\Lambda} e_{10} + \varepsilon e_7^T \Omega e_7 - e_{13}^T \Omega e_{13} + e_8^T G^T G e_8 - e_{11}^T e_{11} + e_9^T H^T H e_9 - e_{12}^T e_{12}, e_i^T = [\mathbf{0}_{n \times (i-1)n} \quad I_{n \times n} \quad \mathbf{0}_{n \times (13-i)n}], \bar{\Pi}_0 = A e_1 + B e_6 + W e_{10} + (1 - \bar{\beta}) K (e_6 + e_{13}) + \bar{\beta} \bar{\alpha} K e_{11} + \bar{\beta} (1 - \alpha) K e_{12}, \bar{\Pi}_1 = \bar{\alpha} K e_{11} + (1 - \bar{\alpha}) K e_{12}, \bar{\Pi}_2 = K e_{11} - K e_{12}, T_1 = [e_6 - e_2; e_1 - e_6], T_2 = [e_8 - e_4; e_1 - e_8], T_3 = [e_9 - e_5; e_1 - e_5], T_4 = [e_7 - e_3; e_1 - e_3].$

Proof. Choose the following Lyapunov function

$$V(x(t)) = \sum_{i=1}^{3} V_i(x(t)), \tag{18}$$

where

$V_1(x(t)) = x^T(t) P x(t),$

$V_2(x(t)) = \int_t^{t-\tau_1} x^T(s) Q_1 x(s) ds + \int_t^{t-d} x^T(s) Q_2 x(s) ds + \int_t^{t-\eta} x^T(s) Q_3 x(s) ds$

$\qquad\qquad + \int_t^{t-\tau_2} x^T(s) Q_4 x(s) ds,$

$V_3(x(t)) = \tau_1 \int_0^{-\tau_1} \int_t^{t+\theta} \dot{x}^T(s) R_1 \dot{x}(s) ds du + d \int_0^{-d} \int_t^{t+\theta} \dot{x}^T(s) R_2 \dot{x}(s) ds du$

$\qquad\qquad + \eta \int_0^{-\eta} \int_t^{t+\theta} \dot{x}^T(s) R_3 \dot{x}(s) ds du + \tau_2 \int_0^{-\tau_2} \int_t^{t+\theta} \dot{x}^T(s) R_4 \dot{x}(s) ds du.$

The derivation of (18) along (11) yields

$$E\{\dot{V}_1(x(t))\} = 2x^T(t)P\bar{\Pi}_0 = \xi^T(t)\left(e_1^T P\bar{\Pi}_0 + \bar{\Pi}_0^T Pe_1\right)\xi(t), \tag{19}$$

$$E\{\dot{V}_2(x(t))\} = \xi^T(t)\left[e_1^T(Q_1 + Q_2 + Q_3 + Q_4)e_1 - e_2^T Q_1 e_2 - e_4^T Q_2 e_4 - e_5^T Q_3 e_5\right.$$
$$\left. - e_3^T Q_4 e_3\right]\xi(t), \tag{20}$$

$$E\{\dot{V}_3(x(t))\} = E\left\{\dot{x}^T(t)\left(\tau_1^2 R_1 + d^2 R_2 + \eta^2 R_3 + \tau_2^2 R_4\right)\dot{x}(t)\right\}$$
$$- \tau_1 \int_t^{t-\tau_1}\dot{x}^T(s)R_1\dot{x}(s)ds - d\int_t^{t-d}\dot{x}^T(s)R_2\dot{x}(s)ds$$
$$- \eta \int_t^{t-\eta}\dot{x}^T(s)R_3\dot{x}(s)ds - \tau_2\int_t^{t-\tau_2}\dot{x}^T(s)R_4\dot{x}(s)ds, \tag{21}$$

where $\xi^T(t) = \begin{bmatrix} x^T(t)\ x^T(t-\tau_1)\ x^T(t-\tau_2)\ x^T(t-d)x^T(t-\eta)x^T(t-\tau_1(t))x^T \end{bmatrix}$
$(t-\tau_2(t))x^T(t-d(t))\ x^T(t-\eta(t))\ \phi^T(Dx(t))\ g(x^T(t-d(t)))\ h(x^T(t-\eta(t)))]$.
Let $R = \tau_1^2 R_1 + d^2 R_2 + \eta^2 R_3 + \tau_2^2 R_4$, we get

$$E\{\dot{x}^T(t)R\dot{x}(t)\} = \xi^T(t)\left\{\bar{\Pi}_0^T R\bar{\Pi}_0 + \rho_2^2\bar{\Pi}_1^T R\bar{\Pi}_1 + (\bar{\beta}^2\rho_1^2 + \rho_1^2\rho_2^2)\bar{\Pi}_2^T R\bar{\Pi}_2\right.$$
$$\left. + \rho_2^2\left(e_7^T + e_{13}^T\right)K^T RK(e_7 + e_{13})\right\}\xi(t). \tag{22}$$

By Jensens inequality and Lemma 1, we obtain

$$- \tau_1 \int_{t-\tau_1}^t \dot{x}^T(s)R_1\dot{x}(s)ds$$
$$= - \tau_1 \int_{t-\tau_1}^{t-\tau_1(t)}\dot{x}^T(s)R_1\dot{x}(s)ds - \tau_1\int_{t-\tau_1(t)}^t\dot{x}^T(s)R_1\dot{x}(s)ds$$
$$\leq - \frac{\tau_1}{\tau_1-\tau_1(t)}\left[x(t-\tau_1(t))-x(t-\tau_1)\right]^T \times R_1 \times \left[x(t-\tau_1(t))-x(t-\tau_1)\right] \tag{23}$$
$$- \frac{\tau_1}{\tau_1(t)}\left[x(t)-x(t-\tau_1(t))\right]^T \times R_1 \times \left[x(t)-x(t-\tau_1(t))\right]$$
$$\leq - \xi^T(t)T_1^T\bar{R}_1^T T_1\xi(t),$$

Similar to (23), we get the following inequalities.

$$d\int_{t-d}^t\dot{x}^T(s)R_2\dot{x}(s)ds \leq -\xi^T(t)T_2^T\bar{R}_2^T T_2\xi(t), \tag{24}$$

$$- \eta\int_{t-\eta}^t \leq -\xi^T(t)T_3^T\bar{R}_3^T T_3\xi(t), \tag{25}$$

$$- \tau_2\int_{t-\tau_2}^t\dot{x}^T(s)R_4\dot{x}(s)ds \leq -\xi^T(t)T_4^T\bar{R}_4^T T_4\xi(t). \tag{26}$$

From Assumption 2, we have

$$x^T(t-d(t))G^T Gx(t-d(t)) - g^T(x(t-d(t)))g(x(t-d(t))) \geq 0, \tag{27}$$

$$x^T(t - \eta(t))H^T Hx(t - \eta(t)) - h^T(x(t - \eta(t)))h(x(t - \eta(t))) \geq 0. \qquad (28)$$

From (4), we get

$$[\phi_i(Dix(t)) - l_i^- Dix(t)]^T[\phi_i(Dix(t)) - l_i^+ Dix(t)] \leq 0, \qquad (29)$$

$$\exists \lambda_i > 0, -2\sum_{i=1}^{m} \lambda_i[\phi_i(Dix(t)) - l_i^- Dix(t)]^T[\phi_i(Dix(t)) - l_i^+ Dix(t)] \geq 0. \qquad (30)$$

Obviously, (30) is equalvalent to

$$-2\Lambda\phi^T(Dx(t))\phi(Dx(t)) - 2x^T(t)\Lambda D^T LDx(t) + 2\phi^T(Dx(t))\Lambda(L^- + L^+)Dx(t) \geq 0. \qquad (31)$$

By combining (9), it is clear that

$$E\{\dot{V}(x(t))\} \leq \xi(t)^T \Xi \xi(t). \qquad (32)$$

If (17) holds, $E\{\dot{V}(x(t))\} \leq 0$ holds. Therefore, the proof is completed.

Theorem 2. *For given delay parameters* $\tau_1 > 0, \tau_2 > 0, d > 0, \eta > 0$ *and constants* $\bar{\alpha} \in [0,1], \bar{\beta} \in [0,1], \epsilon \in [0,1],$ *the closed-loop system* (10) *under cyber-attacks is asymptotically stable with controller gain* $K = YX^{-1}$, *if there exist matrices* $P \in \mathbb{R}^{n \times n} > 0, T_i \in \mathbb{R}^{n \times n} > 0, R_i \in \mathbb{R}^{n \times n}(i = 1, 2, \cdots, 4), \Omega \in \mathbb{R}^{n \times n}, \Lambda = diag\{\lambda_1, \lambda_2, \ldots, \lambda_m\},$ *scalars* $\lambda > 0$ *such that*

$$\bar{R}_i = \begin{pmatrix} R_i & T_i \\ * & R_i^T \end{pmatrix} \geq 0, \qquad (33)$$

$$\begin{bmatrix} \Phi & \bar{\Omega}_1^T & \bar{\Omega}_2^T & \bar{\Omega}_3^T & \bar{\Omega}_4^T & \bar{\Omega}_5^T \\ * & \tilde{R} & \mathbf{0}_{4n \times 4n} & \mathbf{0}_{4n \times 4n} & \mathbf{0}_{4n \times 4n} & \mathbf{0}_{4n \times 4n} \\ * & * & \tilde{R} & \mathbf{0}_{4n \times 4n} & \mathbf{0}_{4n \times 4n} & \mathbf{0}_{4n \times 4n} \\ * & * & * & \tilde{R} & \mathbf{0}_{4n \times 4n} & \mathbf{0}_{4n \times 4n} \\ * & * & * & * & \tilde{R} & \mathbf{0}_{4n \times 4n} \\ * & * & * & * & * & \tilde{R} \end{bmatrix} < 0 \qquad (34)$$

where
$\Phi = e_1^T P\bar{\Pi}_0 + \bar{\Pi}_0^T Pe_1 + e_1^T(Q_1 + Q_2 + Q_3 + Q_4)e_1 - e_2^T Q_1 e_2 - e_4^T Q_2 e_4 - e_5^T Q_3 e_5 - e_3^T Q_4 e_3 - T_0^T \bar{R}_1 T_0 - T_1^T \bar{R}_2 T_1 - T_2^T \bar{R}_3 T_2 - T_3^T \bar{R}_4 T_3 - 2e_{10}^T \Lambda e_{10}$
$+ 2e_{10}^T \Lambda(L^+ + L^-)De_1 - 2e_1^T D^T \Lambda De_1 + \varepsilon e_7^T \bar{\Omega} e_7 - e_{13}^T \bar{\Omega} e_{13} + e_8^T G^T Ge_8 - e_{11}^T e_{11} + e_9^T H^T He_9 - e_{12}^T e_{12}, \tilde{R} = diag\{-2\mu_1 X + \mu_1^2 R_1, -2\mu_2 X + \mu_2^2 R_2, -2\mu_3 X + \mu_3^2 R_3,$
$-2\mu_4 X + \mu_4^2 R_4\},$
$\bar{\Omega} = diag\{X, X, X, X\}\Omega, T_0 = e_7 + e_{13}.$

Proof. By Schur complements, we know that (17) is equalvalent to

$$\begin{bmatrix} \Phi & \Omega_1^T & \Omega_2^T & \Omega_3^T & \Omega_4^T & \Omega_5^T \\ * & -\bar{R}^{-1} & \mathbf{0}_{4n \times 4n} & \mathbf{0}_{4n \times 4n} & \mathbf{0}_{4n \times 4n} & \mathbf{0}_{4n \times 4n} \\ * & * & -\bar{R}^{-1} & \mathbf{0}_{4n \times 4n} & \mathbf{0}_{4n \times 4n} & \mathbf{0}_{4n \times 4n} \\ * & * & * & -\bar{R}^{-1} & \mathbf{0}_{4n \times 4n} & \mathbf{0}_{4n \times 4n} \\ * & * & * & * & -\bar{R}^{-1} & \mathbf{0}_{4n \times 4n} \\ * & * & * & * & * & -\bar{R}^{-1} \end{bmatrix} < 0, \qquad (35)$$

where
$$\Omega_1^T = \begin{bmatrix} \tau_1 \bar{\Pi}_0^T & d\bar{\Pi}_0^T & \eta\bar{\Pi}_0^T & \tau_2\bar{\Pi}_0^T \end{bmatrix}, \Omega_2^T = \begin{bmatrix} \bar{\beta}\rho_1\tau_1\bar{\Pi}_2^T & \bar{\beta}\rho_1 d\bar{\Pi}_2^T & \bar{\beta}\rho_1\eta\bar{\Pi}_2^T & \bar{\beta}\rho_1\tau_2\bar{\Pi}_2^T \end{bmatrix},$$
$$\Omega_3^T = \begin{bmatrix} \rho_1\rho_2\tau_1\bar{\Pi}_2^T & \rho_1\rho_2 d\bar{\Pi}_2^T & \rho_1\rho_2\eta\bar{\Pi}_2^T & \rho_1\rho_2\tau_2\bar{\Pi}_2^T \end{bmatrix},$$
$$\Omega_4^T = \begin{bmatrix} \rho_2\tau_1\bar{\Pi}_2^T & \rho_2 d\bar{\Pi}_2^T & \rho_2\eta\bar{\Pi}_2^T & \rho_2\tau_2\bar{\Pi}_2^T \end{bmatrix},$$
$$\Omega_5^T = \begin{bmatrix} \rho_2(e_7^T + e_3^T)\tau_1 K^T & \rho_2(e_7^T + e_3^T)dK^T & \rho_2(e_7^T + e_3^T)\eta K^T & \rho_2(e_7^T + e_3^T)\tau_2 K^T \end{bmatrix},$$
$$\bar{R}^{-1} = \text{diag}\left\{ R_1^{-1}, R_2^{-1}, R_3^{-1}, R_4^{-1} \right\}.$$

Define $J_1 = \text{diag}\{\underbrace{I, \ldots, I}_{13}, \underbrace{X, \ldots, X}_{20}\}, X = P^{-1}, Y = KX$.

Pre- and post-multiplying (35) with J_1, we get

$$\begin{bmatrix} \Phi & \bar{\Omega}_1^T & \bar{\Omega}_2^T & \bar{\Omega}_3^T & \bar{\Omega}_4^T & \bar{\Omega}_5^T \\ * & -\hat{R} & \mathbf{0}_{4n\times 4n} & \mathbf{0}_{4n\times 4n} & \mathbf{0}_{4n\times 4n} & \mathbf{0}_{4n\times 4n} \\ * & * & -\hat{R} & \mathbf{0}_{4n\times 4n} & \mathbf{0}_{4n\times 4n} & \mathbf{0}_{4n\times 4n} \\ * & * & * & -\hat{R} & \mathbf{0}_{4n\times 4n} & \mathbf{0}_{4n\times 4n} \\ * & * & * & * & -\hat{R} & \mathbf{0}_{4n\times 4n} \\ * & * & * & * & * & -\hat{R} \end{bmatrix} < 0, \tag{36}$$

where $\hat{R} = \text{diag}\left\{ XR_1^{-1}X, XR_2^{-1}X, XR_3^{-1}X, XR_4^{-1}X \right\}, \bar{X} = \text{diag}\{X, X, X, X\}$, $\bar{\Omega}_i = \bar{X}\Omega_i, i = 1, \cdots, 5$.

From Lemma 2, it follows that for $\forall i = 1, \cdots, 4, -XR_i^{-1}X \le -2\epsilon_i X + \epsilon_i^2 R_i$. Substituting $-2\epsilon_i X + \epsilon_i^2 R_i$ for $-XR_i^{-1}X$, we derive that (36) is equalvalent to (34). The proof is completed.

We propose the following algorithm to solve BMIs (33)–(34).

Algorithm 1

Step 1: Input system matrices;
Step 2: Select parameters τ_1, τ_2, d, η;
Step 3: Solving the BMIs (33)-(34) in Theorem 2 by Matlab LMI toolbox;
    if $tmin < 0$
        a feasible solution is found;
        break;
    else
        go back to Step 2;
    end if
Step 4: Output the gain matrix K.

---

4 Numerical Examples

In this section, a numerical example will be given to illustrate the feasibility of the proposed control method for event-triggered synchronization of chaotic delayed Lur'e systems with random cyber-attacks.

For system (1) and (3), the state matrices are as follows:

$$A = \begin{bmatrix} -\frac{8}{7} & 9 & 0 \\ -5 & -0.914 & 1 \\ 0 & 12 & -0.8 \end{bmatrix}, B = \begin{bmatrix} -0.1 & 0 & 0 \\ -0.1 & 0 & 0 \\ 0.2 & 0 & -0.1 \end{bmatrix}, W = \begin{bmatrix} \frac{27}{7} & 0 & 0 \\ 0 & 0 & 0 \\ 0 & 0 & 0 \end{bmatrix}.$$

The nonlinear characteristic of the system is chosen as $\sigma_i(x) = \frac{1}{2}(|x+1| - |x-1|), i = 1, 2, 3$.

The event-triggered weighting matrix is determined as

$$\Omega = \begin{bmatrix} 19.9175 & -2.9971 & 7.2135 \\ -2.9971 & 30.6072 & -1.9185 \\ 7.2135 & -1.9185 & 11.4720 \end{bmatrix}.$$

Set $\tau_1 = 0.4, \tau_2 = 0.4, d = 0.7, \eta = 0.8$, the initial states of the master system and slave systems $m(0) = [0.5\ 0.5\ 1]^T$ and $s(0) = [0.5\ 1.5\ 1]^T$. Then, the initial states of the error system are $x(0) = m(0) - s(0) = [1\ -1\ 2]^T$.

In the following two cases, the selection of the above given parameters is the same. Section 4.1 is used to illustrate that the controller designed can keep the systems synchronized for both event-triggered master and slave systems without cyber-attacks. Section 4.2 shows that the controller designed can also achieve this function for the discussed chaotic delayed Lur'e systems even when the system is attacked.

4.1   Design Method for Synchronization Error Systems Without Cyber-Attacks

Set $\beta = 0$, it signifies that event-triggered chaotic delayed Lur'e systems work regularly without cyber-attacks.

By solving LMIs (34) based on LMI toolbox in MATLAB, we compute the controller gain matrix as

$$K = \begin{bmatrix} -0.0028 & -0.7319 & -0.0043 \\ -0.1208 & -0.0166 & -0.1012 \\ -0.0918 & 1.1955 & -0.1225 \end{bmatrix}.$$

Under the corresponding designed gain matrix, we get Figs. 2, 3 and 4. Figure 2 presents the state response of $x(t)$ and the controller $u(t)$ is shown in Fig. 3. We can find that event-triggered distributed control design approach is useful to ensure the event-triggered synchronization of the master system and slave system without cyber-attacks. Based on the event-triggered method (7), the event-triggered instants of $x(t)$ are shown in Fig. 4.

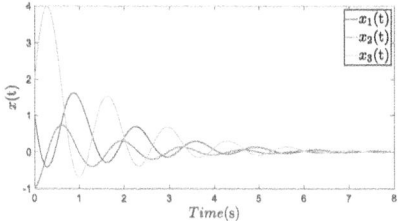

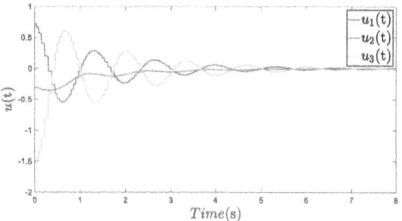

Fig. 2. Response of error signal $x(t)$ for synchronization error systems without cyber-attacks

Fig. 3. Response of control input $u(t)$ for synchronization error systems without cyber-attacks

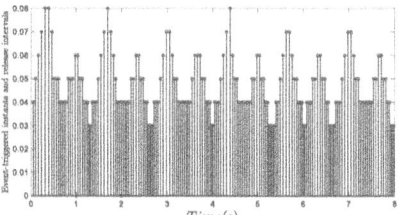

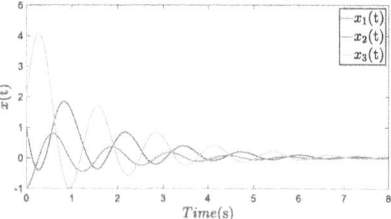

Fig. 4. Event-triggered instants and release intervals

Fig. 5. Response of error signal $x(t)$ for synchronization error systems with cyber-attacks

4.2 Design Method for Synchronization Error Systems with Cyber-Attacks

Set $\bar{\beta} = 0.5, \bar{\alpha} = 0.5$, which means that the system is subject to random cyber-attacks with the probability of 50%, and the probability of switching between two different types of network attacks is 50%. Using non-linear functions $g(x(t))$ and $h(x(t))$ to represent two types of cyber-attacks as follows $g(x(t)) = h(x(t)) = \text{diag}\{0.1\tanh(x_1(t)), 0.2\tanh(x_2(t)), 0.23\tanh(x_3(t))\}$. It can be seen that $g(x(t))$ and $h(x(t))$ satisfy Assumption 2 with $G = H = diag\{0.1, 0.2, 0.23\}$. The state response $x(t)$ and the controller $u(t)$ of the event-triggered chaotic delayed Lur'e systems with cyber-attacks are shown in Fig. 5 and Fig. 6, respectively. The switching rule diagram between the two types of cyber-attacks is shown in Fig. 7. The graph of the existence of cyber-attacks is shown in Fig. 8. Based on the event-triggered method (7), the event-triggered instants of error states $x(t)$ are shown in Fig. 9. We can observe that even if the system is attacked by random network, the event-triggered master and slave systems can still synchronize due to the role of the controller. Compare Fig. 2 with Fig. 5, the state response $x(t)$ of the error system subject to cyber-attacks is almost the same as that without cyber-attacks under the controller. It proves that the controller designed method is an effective way to realize the synchronization of the discussed system even with random cyber-attacks.

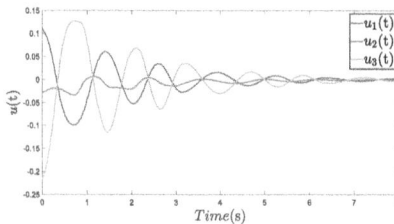

Fig. 6. Response of control input $u(t)$ for synchronization error systems with cyber-attacks

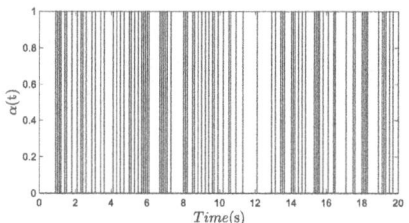

Fig. 7. Trajectory of switching rule $\alpha(t)$ between the cyber-attacks

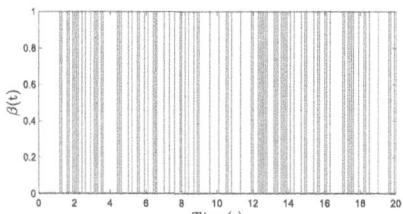

Fig. 8. Trajectory of cyber-attacks $\beta(t)$

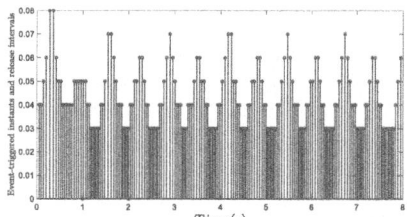

Fig. 9. Event-triggered instants and release intervals

5   Conclusion

This paper investigated the distributed event-triggered synchronization of chaotic delayed Lur'e system with stochastic cyber-attacks. In order to reduce the load of communication network and maintain performance of network, an event-triggered scheme has been introduced. Then, an event-triggered synchronization controller has been constructed. Based on Lyapunov-functional approach, the LMI-based sufficient conditions have been derived to design controllers. Finally, the effectiveness of the proposed controller design method has been demonstrated by a numerical example.

References

1. Lam, H.K., Leung, F.H.F.: Design and stabilization of sampled-data neural-network-based control systems. IEEE Trans. Syst. Man Cybern. Part B (Cybernetics) **36**(5), 995–1005 (2006)
2. Zhao, Y., Small, M.: Minimum description length criterion for modeling of chaotic attractors with multilayer perceptron networks. IEEE Trans. Circuits Syst. I Regul. Pap. **53**(3), 722–732 (2006)
3. Moini, S., Alizadeh, B., Emad, M., Ebrahimpour, R.: A resource-limited hardware accelerator for convolutional neural networks in embedded vision applications. IEEE Trans. Circuits Syst. II Express Briefs **64**(10), 1217–1221 (2017)

4. Carroll, T.L., Pecora, L.M.: Synchronizing chaotic circuits. IEEE Trans. Circ. Syst. **38**(4), 453–456 (1991)
5. Yalçin, M.E., Suykens, J.A.K., Vandewalle, J.: Master-slave synchronization of lur'e systems with time-delay. Int. J. Bifurcat. Chaos **11**(06), 1707–1722 (2001)
6. Liu, S., Wei, G., Song, Y., Liu, Y.: Extended kalman filtering for stochastic nonlinear systems with randomly occurring cyber attacks. Neurocomputing **207**, 708–716 (2016)
7. Albert, A.: Comparison of event-triggered and time-triggered concepts with regard to distributed control systems. Embedded world **235–252**, 2004 (2004)
8. Yue, D., Tian, E., Han, Q.L.: A delay system method for designing event-triggered controllers of networked control systems. IEEE Trans. Autom. Control **58**(2), 475–481 (2013)
9. Liu, J., Wei, L., Xie, X., Tian, E., Fei, S.: Quantized stabilization for t-s fuzzy systems with hybrid-triggered mechanism and stochastic cyber-attacks. IEEE Trans. Fuzzy Syst. **26**(6), 3820–3834 (2018)
10. Lang, J., Zhang, Y., Zhang, B.: Event-triggered network-based synchronization of delayed neural networks. Neurocomputing **190**, 155–164 (2016)
11. Hua, C., Ge, C., Guan, X.: Synchronization of chaotic lur'e systems with time delays using sampled-data control. IEEE Trans. Neural Netw. Learn. Syst. **26**(6), 1214–1221 (2014)
12. Lee, S.H., Park, M.J., Kwon, O.M.: Synchronization criteria for delayed lur'e systems and randomly occurring sampled-data controller gain. Commun. Nonlinear Sci. Numer. Simul. **68**, 203–219 (2019)
13. Wang, Y., Zhu, Y., Zhu, Y.: On aperiodic event-triggered master-slave synchronization of chaotic lur'e systems. J. Franklin Inst. **356**(17), 10576–10592 (2019)
14. Gu, Z., Yan, S., Park, J.H., Xie, X.: Event-triggered synchronization of chaotic lur'e systems via memory-based triggering approach. IEEE Trans. Circuits Syst. II Express Briefs **69**(3), 1427–1431 (2021)
15. Ni, Y., Wang, Z., Fan, Y., Huang, X., Shen, H.: Memory-based event-triggered control for global synchronization of chaotic lur'e systems and its application. IEEE Trans. Syst. Man Cybern. Syst. **53**(3), 1920–1931 (2023)
16. Ding, D., Wang, Z., Ho, D.W., Wei, G.: Observer-based event-triggering consensus control for multiagent systems with lossy sensors and cyber-attacks. IEEE Trans. Cybern. **47**(8), 1936–1947 (2016)
17. Park, P., Ko, J.W., Jeong, C.: Reciprocally convex approach to stability of systems with time-varying delays. Automatica **47**(1), 235–238 (2011)
18. Xiong, J., Lam, J.: Stabilization of networked control systems with a logic ZOH. IEEE Trans. Autom. Control **54**(2), 358–363 (2009)

In-Game Win Prediction Models for Cricket

Sonish Lamsal$^{(\boxtimes)}$ and David Kahle

Baylor University, Waco, TX 76706, USA
`sonish_lamsal1@baylor.edu`

Abstract. The growth of cricket as a global sport cannot be understated. Indeed, in recent years cricket has experienced the most rapid growth rates on many business and popularity measures, both domestic and global. However, cricket as sport is surprisingly understudied and thus presents exciting opportunities for advanced analytics. Building on the dynamic logistic regression framework of Asif and McHale (2016), this paper introduces a improved methodology for in-game win prediction in the context of Indian Premier League (IPL) T20-style cricket matches using ball-by-ball data. After transitioning the model to a Bayesian setting, the work makes two specific advances: integrating historical data via power priors and more transparent and flexible smoothing of the dynamic coefficients via Gaussian processes. To assess the performance of the proposed method relative to that of Asif and McHale, we use cross validation on data from hundreds of IPL matches. Metrics such as the Brier score and accuracy suggest that the proposed method exhibits considerably better predictive performance in addition to the increased optionality afforded to the modeler.

Keywords: Sports analytics · cricket · in-game win prediction · Bayesian statistics · dynamic logistic regression · power priors · Gaussian processes

1 Introduction

Cricket, one of the most widely played and followed sports in the world, has captivated millions with its unique blend of strategy, skill, and uncertainty. However, cricket has received relatively limited attention in terms of published research on forecasting. Most of the existing research on cricket forecasting has focused on pre-match predictions. However, with the increasing popularity of in-play betting across all sports, which allows bettors to place bets during a game or match, there is a growing demand for models that can provide forecasts in real-time as the game progresses. Cricket in particular is well-suited for in-play betting due to its discrete nature. Unlike sports like soccer, cricket's discrete gameplay offers numerous opportunities for bookmakers and bettors to actively participate in the markets during the game. Consequently, cricket attracts significant volumes of in-play betting. Asif and McHale [2] report that the total amount bet during a typical major One-Day International (ODI) involving Pakistan or India is on the order of $1bn. In this paper, we present a Bayesian modification to the in-play forecasting model proposed by Asif and McHale [2] for Indian Premier League (IPL) cricket, which involves a match format that is shorter than ODIs called Twenty20 (T20). The proposed model can be used to estimate the probability of victory for a team at any given moment during a cricket match.

Cricket is a bat-and-ball game involving two teams of eleven players who take turns batting and bowling on a field centered around a 22-yard pitch. The objective is for the team batting first to score as many runs as possible, setting a target for the team batting second. The second team then attempts to meet or surpass this target. The team with the most runs at the end of the game wins the match.

Here are some specific terms that are relevant to our discourse:

Wicket. Wicket can refer to either (1) the physical structure comprised of three stumps and two bails that the bowler aims to hit or (2) a dismissal, where a batsman is ruled out in various ways, including being bowled, caught, run out, or leg before wicket (LBW).

Ball. Ball is the term for both (1) the central object used in the game and (2) the delivery bowled by the bowler to the batsman, analogous to a pitch in baseball. A standard cricket ball is made of cork covered by leather, traditionally red in test matches and white in limited-overs cricket. This is similar to a pitch in baseball.

Over. Over consists of six legal deliveries bowled by a bowler. After each over, a different bowler from the opposite end of the pitch bowls the next over, and play continues this way throughout the innings.

Innings. Innings refers to a period in which a team takes its turn to bat. The term is both singular and plural, in contrast to baseball's inning (sing.) and innings (pl.). In different formats, teams have either one or two innings to score runs. An innings for a team comes to an end if they lose all their wickets or there are no more overs left. One half of an inning in baseball is comparable to a single innings in cricket.

Runs. Runs are the basic unit of scoring in cricket. Batsmen score runs by hitting the ball and running to the opposite end of the pitch, exchanging ends with the non-striker to earn a single run. Batsmen can also score runs by hitting the ball out of the playing field's boundaries. When the ball crosses these boundary lines without touching the ground, the batting team earns six runs (also called maximum). If the ball reaches the boundary after touching the ground within the field of play, the batting team earns four runs. This is referred to as a "four" or a "boundary."

Cricket is played in three main formats, each with its own set of rules and game timing. In the T20 format, each team bats for a maximum of 20 overs, making for a fast-paced and dynamic game. One-day internationals (ODIs) extend the game, with each team batting for 50 overs, blending strategic depth with the excitement of limited-overs cricket. The third format, test matches (tests), represent the game's traditional format, with no limit on the number of overs and the game stretching over five days, allowing for two innings per team. This diversity in formats makes cricket a versatile and globally popular sport, appealing to fans of all preferences.

Two of the most important works in the field of cricket analytics is the pair of works Duckworth and Lewis [9,10]. These works are so widely recognized that they have been adopted to reset targets in official cricket matches. In cricket, targets are reset in limited-overs matches (like ODIs and T20s) when the match is interrupted by rain or any other unforeseen circumstances, potentially altering the number of overs each team can play. The Duckworth-Lewis-Stern method is a mathematical formula used to calculate revised targets in these scenarios, ensuring a fair outcome by considering the number of overs and wickets in hand at the time of interruption. It adjusts the target based on the resources (overs and wickets) remaining, ensuring that the revised target reflects the scoring potential had the interruption not occurred. This method aims to level the playing field, taking into account the efforts of both teams before and after any disruptions.

Most of the relevant works in cricket analytics have focused on pre-match win predictions. Brooks et al. [6] estimated test match outcome probabilities using an ordered response model using a loss, draw, win scale. They note that the outcome of a test match can be explained through the contributions of batting and bowling efforts. Scarf et al. [27] developed a multinomial logistic regression model to forecast match outcome probabilities. At the beginning of each session, the probabilities of match outcomes (win, lose, draw) are predicted through a series of multinomial logistic regression models. They explore the session-by-session variations in the probabilities of different match outcomes and the effects of associated covariates with the objective of assisting team management in determining the most appropriate time to declare during an innings. In 2012, Akhtar and Scarf [1] introduced a multinomial logistic regression model aimed at predicting test match outcomes, providing team captains and management with analytical support to strategize batting approaches-be they aggressive or defensive—for upcoming sessions. Additionally, Bailey and Clarke [3] developed a forecasting model specifically designed to predict the margin of victory in limited overs cricket.

There has been a growing trend towards the use of Bayesian methods in sports analytics. The appeal of these techniques lies in their ability to incorporate expert knowledge and pre-existing beliefs, utilize Bayesian updating to refine these beliefs with fresh data, and adeptly handle complex problems, as highlighted by Santos-Fernandez et al. [26]. Efron and Morris [11] used an empirical Bayes approach to predict baseball batting averages. Similarly, Neal et al. [22] employed empirical Bayesian methods to predict batting averages in the second half of the season based on the first half. Other notable examples of using empirical Bayesian approach to predict baseball include Brown [7] and Jiang and Zhang [17]. Jensen et al. [16] employed a Bayesian hierarchical model to make projections on the number of home runs per season for players. In soccer, Karlis and Ntzoufras [19] preformed a Bayesian modelling of Poisson difference distribution to capture the disparity in goals scored during football matches, utilizing data from the English Premier League. Suzuki et al. [30] proposed a Bayesian model that factored in the FIFA World Ranking to predict the outcomes of the 2006 World Cup.

Following the trend of using Bayesian techniques in sports, cricket analytics is also beginning to embrace these methods. Boys and Philipson [4] employed a Bayesian approach to rank test cricketers across different eras by modeling the natural variation in their batting scores since the first test in 1877, using an additive log-linear model that accounts for year, age, and cricket-specific factors. Stevenson and Brewer [29] introduced a Bayesian survival analysis method to predict international cricketers' batting abilities and analyzed performance improvements from the start of their innings. They then used this model to evaluate New Zealand's opening batsmen, influencing team selection and batting order strategies. Expanding on the effectiveness of the Bayesian approach in sports analytics, in this article we use Bayesian methods to make in-game win predictions in cricket in this paper.

More specifically in this paper we present an Bayesian in-game probability forecasting model for T20 games in this paper. The primary objective of the work is to construct a Bayesian version of Asif and McHale's model capable of incorporating historical data. While most of the aspects remain similar, we propose using power prior to incorporate the prior information we have available and use Gaussian processes to smooth the regression coefficients. The model is similar to Asif and McHale in the sense that, it too allows the parameters (the coefficients of covariates) to evolve smoothly as the match progresses. We first introduce the data and the describe the covariates in Sect. 2. Then, we talk about the model by Asif and McHale in Sect. 3 along with our proposed model in Sect. 4. Section 6 has the application of the model into IPL data along with the comparison of the model performances.

2   Data

We obtained ball-by-ball IPL data for all the games played over the 2011 to 2020 seasons from Kaggle [18]. Matches with incomplete data or interrupted play (i.e. no results) were excluded from our analysis. For historical data, we collected ball-by-ball data for T20 matches between club and countries from 2005 to 2010 from cricsheet [8]. The data from these set of games is incorporated in our model with the help of power priors as discussed in Sect. 4.1.

We categorize variables into two groups: pre-match covariates, which are measured before the match begins, and in-play covariates, which are exclusively measured during the match. Pre-match covariates are the ones that would be known before the game starts, e.g. home venue, and remain fixed throughout the game. In-play covariates are the ones that would come into play once the game starts, e.g. number of wickets lost, and change as the game progresses. The following section contains a further description about the covariates.

2.1   Pre-match Covariates

Pre-match covariates refer to variables that can be measured before the game starts. Several factors may influence the likelihood of a match outcome before the match begins, for example home venue, winning the toss to determine batting order, the impact of day-night conditions, team quality, and the team's current form.

It is a general belief in cricket as in most sports that teams playing at their own venue enjoy certain advantages. Morley and Thomas [20] investigate the home-field effect with respect to English county cricket and find that home teams won 57% of matches with a win/loss result. They also conduct a literature review that concludes "In general the evidence indicates clear, and historically stable, home-field advantage although its magnitude varies between sports and, to an extent, between different competitions and levels within a sport..." Several explanations are supposed relevant, for example extensive experience playing at the venue, the presence of the supportive audience, and traveling fatigue. Whatever the case, winning the toss in cricket is widely regarded as advantageous for a team. However, the literature is not entirely one-sided on the topic, and at least some previous studies have not found statistical significance in the effect of the binary covariate of winning the toss on the match outcome [1]. Like Asif and McHale [2], we also did not find toss to be an important factor in predicting the match outcome in both the first and second innings.

The relative ability of the two teams is also known, albeit with uncertainty, in advance of the match through ratings and recent performance trends, and it is appropriate to use this information in predicting which will win. Along these lines, Asif and McHale [2] use ODI ratings from the International Cricket Council (ICC) to create a ratings difference variable to use as a pre-match covariate. Unlike that setting, however, the ICC does not provide official ratings for Indian Premier League (IPL) teams. Consequently, we adopted the Elo rating system to evaluate the teams [12, 13]. The Elo rating system is a long-established, simple method for assessing the relative skill levels of players or (as in this case) teams in various competitive environments. Initially designed for two-player games such as chess, it applies equally to any system that involves two-player contests. In our context, the players are the teams.

The basic idea of the Elo system is to attribute to each team a performance distribution from which draws are taken when a contest takes place and the team with the higher performance wins. The distributions are assumed to be the same up to location shift, typically normal, with the locations (means) of the distributions representing the team's true skill. After a contest takes place, both teams locations are updated conditional on the result. The Elo rating system calculates the expected score for each team before a match using the formula

$$E_A = \frac{1}{1 + 10^{(R_B - R_A)/D}}, \quad E_B = \frac{1}{1 + 10^{(R_A - R_B)/D}} \quad (1)$$

Here, E_A and E_B represent the expected scores for Team A and Team B, respectively, while R_A and R_B denote their current rating or score and D is a divisor that adjusts the sensitivity. Every team is given the same starting score and the rating gets updated as they play their games. This formula estimates each team's probability of winning, with the team having a higher rating being more likely to win. After the match, the teams' observed scores are used to adjust their ratings. The adjustment is done using the formula $R' = R + K(S - E)$, where R' is the new rating, R the previous rating, and K is the "K-factor," which plays a crucial role in the sensitivity of the rating adjustment. The K-factor determines how much a single game can affect the rating, with a higher K-factor causing larger changes in ratings after each game. This factor can vary based on the level of play or the importance of the match, with typical values ranging for established teams. S is the actual score of the match (1 for a win, 0 for a loss, and 0.5 for a draw), and E is the expected score determined before the match. As discussed by Nishanth [23], Elo rating can be a good tool to rank IPL teams. Figure 1, generated by Nishanth [23], depicts the evolution of the Elo scores of all IPL teams over the years, showing their relative skills; each dot represents the score at the end of each season. We used initial value of 1500 for each team and a value of 400 for the sensitivity parameter (D) in our calculations which was done in R Core Team [24] using `elo` package [14].

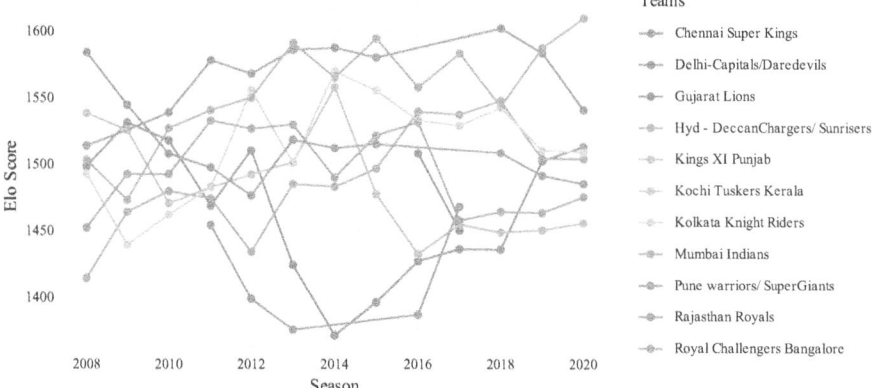

Fig. 1. Ranking Of Teams Using Elo Over the Years.

Similar to the method used by Asif and McHale, we assess a team's current performance by employing a weighted average of their outcomes (win or loss) from their last five matches. This approach helps us gauge the level of "good form" a team has displayed over their recent matches. A team achieving victories in all of their past five games achieves a perfect form score of 1. Conversely, a form score of 0 is allocated to a team that has failed to secure any wins in those matches.

2.2   In-Play Covariates

In-play covariates capture the dynamic state of play or team position as the innings progress. The current state of a cricket match can be simplified to three key elements: runs scored (or runs needed to win in the second innings), wickets lost, and the number of remaining balls (or overs). Our model includes two variables to account for runs. In the first innings, we use the run rate, or runs per over (RPO). In the second innings, the focus shifts to the required runs per over (RRPO) for the batting team to secure victory.

The raw number of wickets lost does not always adequately reflect the urgency of a teams predicament. The value of losing a wicket depends on factors such as which wicket is lost and at what stage of the match. Following Asif and McHale [2], we employ a measure called wicket resources lost (WRL) instead of simply considering the number of wickets lost in an attempt to acknowledge the varying importance of wickets in cricket. Teams typically place stronger batsmen at the top of the order, and the significance of each wicket changes throughout the innings. Towards the end of the game, losing a wicket will have larger effect on the batting team's total if they are low on wicket than if they had only lost one or two wicket. For example, losing a wicket with 10 overs remaining will have greater effect on a team's ability to score if they have 6 wickets in hand in comparison to having 8 wickets in hand. The impact on the team's resources will differ based on these considerations. As discussed in Sect. 1, Duckworth and Lewis [9] developed a method for resetting targets in interrupted matches, where they model the expected remaining runs (Z) as a function of the remaining overs (u) and wickets lost (WL). We use this method to calculate our WRL variable. Figure 2 illustrates the evolution of the relationship between WRL and WL as an innings unfolds. We can see the non-linear relationship between the two variable at later stages of the inning compared to the beginning stages.

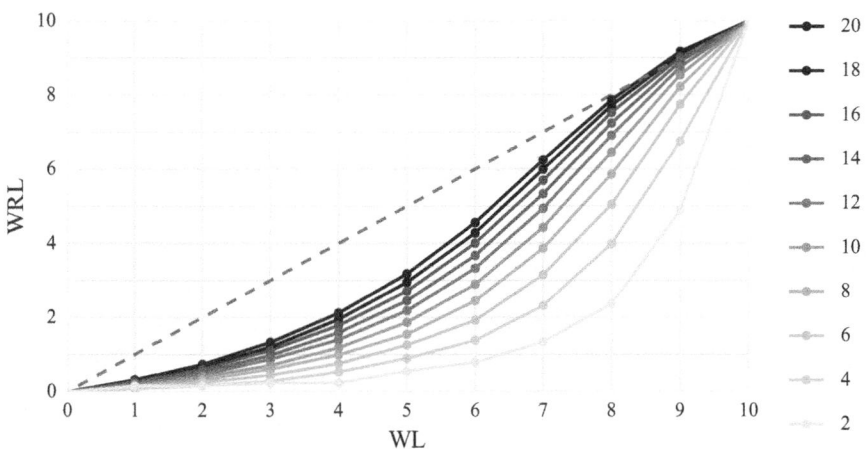

Fig. 2. Relationship between wickets lost (WL) and wicket resources lost (WRL) for $20, 18, \ldots, 2$ (top to bottom) overs remaining.

We adopt the same formula Asif and McHale had to calculate WRL, i.e. $\text{WRL} = \frac{Z(u,0)-Z(u,w)}{Z(u,0)}$, where $Z(u,w)$ is the expected runs in the remainder of the innings with u overs to go and w wickets lost. The expected runs comes form the method described by Duckworth and Lewis [9]. We conclude this section with Table 1, which provides a brief summary of all the variables in use.

Table 1. Variable used and their description.

Variables	Type	Description
home	Pre-match	Binary; whether the team is playing at home
away	Pre-match	Binary; whether the team is playing away
FD	Pre-match	Continuous; the difference between the forms of the reference and opposition teams
RD	Pre-match	Continuous; the difference in the Elo ratings of the reference and opposition teams before the match
toss	Pre-match	Binary; whether the team won the toss
WL	In-play	Discrete; the number of wickets lost
WRL	In-play	Continuous; function of wickets lost and overs remaining
RPO	In-play	Discrete; runs per over scored by the batting team
RRPO	In-play	Discrete; runs per over required in order for the batting (chasing) team to win the match (Only for second innings)

2.3   More on Data

In cricket time is denominated in balls remaining in an innings. The complete ball-by-ball data is structured into a sequence of data matrices, with one matrix corresponding to each ball of each innings, similar to Asif and McHale [2]. Using k to denote the number of balls remaining in the innings and $n(k)$ the number of observations (games) with k balls remaining,[1] each data matrix comprises an observed $n(k) \times M$ vector of binary response variables y_k and a matrix of independent variables denoted by X_k. Not to be confused with the Elo K-factor K, the value of K in our case–the maximum number of balls remaining–is 120 since we are applying this method to T20 games instead of 300 as in sif and McHale [2]. Similar to Asif and McHale, we also observe that matrices may vary in the number of rows (sample size $n(k)$), as not all innings conclude with the entirety of the pre-allotted overs being played. The sample sizes decrease more rapidly for the team batting second than for that batting first, because the second innings can end in more ways than the first innings: either the batting team uses all of its wickets or overs (as in the first innings), or it achieves the runs target [2]. A snippet of the data is provided in Table 2.

3   The Asif and McHale Model

In this section, we explain the methodology proposed by Asif and McHale [2] who adopt a logistic regression model to estimate the probability of the batting team winning the match. They develop two different forecasting models, one for each innings of ODI cricket. They argue this is done because the team batting in

---

[1] Cricket games can end before the number of balls remaining expires. Thus, the number of games observed can decrease as k decreases.

Table 2. Data snippet for second innings at K = 60 balls remaining.

win_loss	home	away	rd	fd	toss	wl	wrl	rpo	rrpo
0	0	1	10.78	−0.00	1	2	0.41	9.10	12.40
0	0	1	100.76	−0.47	1	2	0.41	7.00	8.70
0	0	1	121.95	−0.20	0	3	0.70	8.40	8.10
1	0	1	3.22	−0.27	0	1	0.16	9.00	8.30
0	0	1	−70.26	−0.33	0	5	1.87	8.00	9.40
1	1	0	38.15	0.33	1	3	0.70	5.80	9.60
1	0	1	96.08	−0.40	1	4	1.18	5.00	9.00
0	0	1	45.63	0.07	1	1	0.16	8.70	8.20
0	1	0	8.13	0.87	1	2	0.41	9.30	7.60
1	1	0	8.02	0.40	1	1	0.16	8.00	9.90

the first innings plays with the aim of scoring as many runs as possible, in order to maximize its chances of winning. The team batting in the second innings, on the other hand, plays with the aim of achieving its target before either all of its wickets have been lost or all of its pre-allotted overs have been played. Also, certain covariates, such as the number of runs needed per remaining over (RRPO) to win, are unique to the second innings.

3.1   Model

Like Asif and McHale [2], we adopt a logistic regression model to make in-game forecasts. As the structure of the model varies over the match it is referred to as dynamic logistic regression (DLR) model. It is dynamic in the sense that the parameters of the models, the coefficients, are allowed to vary as the match progresses. The model is being catered to team batting in the innings. So, the model is predicting win probability for the team batting first in the first innings and for the team batting second in the second innings.

Let p_k to represent the probability of the batting team winning the match with k balls remaining, where k ranges from K to 1. So, For each k, the model is therefore

$$\text{logit}(p_k) = \mathbf{X}_k \boldsymbol{\beta}_k, \tag{2}$$

where $\mathbf{X}_k$ represents a design matrix with $M+1$ columns, including M covariates and a vector of ones representing the intercepts and n_k rows representing all the data for k ball remaining, while $\boldsymbol{\beta}_k$ is a vector containing regression parameters.

The Asif and McHale [2] fits K logistic models for each inning, resulting in different estimated coefficients $\boldsymbol{\beta}_k = [\widehat{\beta}_m(k)]$ for each value of k, where $m = 0, \ldots, M$. These estimated coefficients can be thought of as $M+1$ functions $f_m(k|\boldsymbol{\alpha}_m)$ over the range $k = K, K-1, \ldots, 1$:

$$\widehat{\beta}_m(k) = f_m(k|\boldsymbol{\alpha}_m), \tag{3}$$

where $\boldsymbol{\alpha}_m$ is a vector of parameters. Since the logistic models are fit ball-by-ball independently, the estimates do not result in "smooth" curves. In reality they are of course functions over a discrete space, so they cannot be smooth in the traditional sense, but they are nevertheless quite noisy when plotted and therefore smoothed as discussed below.

3.2 Smoothing the Coefficients

After fitting the series of independent logistic models, Asif and McHale proceed to refine their dynamic logistic regression model by smoothing the estimated coefficients for each covariate. For a given covariate m (where $m = 0, 1, \ldots, M$), and at k balls remaining, the estimated coefficient $\widehat{\beta}_m(k)$ is approximated by a smooth function, represented as $\widehat{\beta}_m(k) = f_m(k, \boldsymbol{\alpha}_m)$. In their setup, the selection of f_m is based on an analysis of the scatterplot of the series of parameter estimates $\widehat{\beta}_m(k)$ for each m and involves a subjective assessment and statistical testing to identify the most suitable functional form. They recommend using polynomials of varying degrees. They also propose splines as perhaps another alternatives.

The parameters $\boldsymbol{\alpha}_m$ are estimated by minimizing a weighted sum of squares given by $WSSE(m) = \sum_{k=1}^{K-1} \left(\frac{\widehat{\beta}_m(k) - f_m(k, \boldsymbol{\alpha}_m)}{s.e.(\widehat{\beta}_m(k))} \right)^2$ where $s.e.(\widehat{\beta}_m(k))$ is the standard error of estimated coefficients. They note that the $M+1$ parameters are not estimated independently for any k-balls-remaining logistic model. Thus, after the smoothing first sets of coefficients, the series of independent logistic models was refitted, subject to the parameter constraint $\beta_1(k) = f_1(k, \widehat{\boldsymbol{\alpha}}_1)$. After the estimates of the remaining M parameters were updated for each k, the next series of estimates, $\widehat{\beta}_2(k), \ldots, \widehat{\beta}_2(1)$, was subsequently smoothed in a similar manner. This process is continued until the estimated intercept terms, $\widehat{\beta}_0(k), \ldots, \widehat{\beta}_0(1)$, have been smoothed. The same procedure is then followed for the second innings model.

3.3 Variable Selection

Prior to fitting the final series of k independent logistic models for an innings, the model determines an optimal subset of candidate covariates for their model. They note there are several methods for this: Akaike information criterion (AIC), delete-d cross-validation (CVd) with random subsamples of size $d = n(1 - 1/(\log(n) - 1))$, and k-fold cross validation. They recommend the delete-d cross-validation method since the aim for the model was to develop a model with the maximum forecasting power.

For identifying the best set of independent logistic models during the first innings, the researchers employ the `bestglm()` function in R for each $k = K, \ldots, 1$, incorporating all potential covariates. The list of candidate covariates considered for the DLR model includes the pre-match variables of home, away, toss, DN (day-night), FD (form difference), and RD (rating difference, as determined by Elo in our case) as well as the in-play covariates WRL (wicket resources lost), RPO (runs per over), and all possible two-factor interactions among the categorical variables, e.g. DNT = DN × toss or an interaction term between WRL and RPO. In the second innings, the covariate RPO is substituted with RRPO (required runs per over). One thing to note is that we couldn't use in this paper was the DN(day-night) variable since the IPL games cannot be classified as being a day-night game or not. The frequency with which each covariate appears in the optimal logistic model for both the first and second innings is noted and any variable that appears at least once is included in the model. The results of variable selection is discussed in Sect. 6.1.

4   The Proposed Bayesian Model

The primary objective of this paper is construct a Bayesian version of the Asif and McHale model that is more extensible and incorporates past data. The fundamental structure of the model remains the same in the sense that we also fit a series of logistic regressions for each ball bowled. We use the same variable selection as Asif and McHale [2] described above. The two basic differences are 1) that once the relevant covariates are determined, past data is incorporated via power priors and 2) smoothing is accomplished via Gaussian priors. All of this is done in a Bayesian context. In this section we describe these in more detail.

4.1   Power Priors

Power priors are prior distributions used in Bayesian data analyses that provide a way to incorporate prior data, typically in the form of past studies, into a Bayesian model. The key idea behind power priors is to adjust the strength of the prior based on the relevance–perceived or estimated–of the available data. This is done with the help of a special parameter called the power parameter.

Let $\mathbf{D}$ denote the (current) data and $L(\boldsymbol{\theta}|\mathbf{D})$ its corresponding likelihood, where $\boldsymbol{\theta}$ is a notational device to refer to all the parameters in the model. Similarly, let the historical data be denoted $\mathbf{D}_0$ from a similar previous study with likelihood $L(\boldsymbol{\theta}|\mathbf{D}_0)$. The basic formulation of power priors as discussed by Ibrahim et al. [15] is given by

$$\pi(\boldsymbol{\theta}|\mathbf{D}_0, a_0) \propto L(\boldsymbol{\theta}|\mathbf{D}_0)^{a_0} \pi_0(\boldsymbol{\theta}),$$

where a_0 is the power parameter, a number in the interval $a_0 \in [0, 1]$, and $\pi_0(\boldsymbol{\theta})$ is the prior distribution of the previous study. The power prior is therefore the posterior we obtain using the past data serves along with the prior π_0. Using the power prior above, the corresponding posterior distribution of $\boldsymbol{\theta}$ in light of the past and current data is given by

$$\pi(\boldsymbol{\theta}|\mathbf{D}, \mathbf{D}_0, a_0) \propto L(\boldsymbol{\theta}|\mathbf{D})\pi(\boldsymbol{\theta}|\mathbf{D}_0, a_0) = L(\boldsymbol{\theta}|\mathbf{D})L(\boldsymbol{\theta}|\mathbf{D}_0)^{a_0} \pi_0(\boldsymbol{\theta}).$$

Here, $\pi_0(\boldsymbol{\theta})$ is the prior distribution for the past historical data.

The power prior parameter α_0 controls the degree to which the prior information influences the posterior distribution. Loosely speaking, the value of α_0 determines the proportion of information in the previous data to use, with different values producing different levels of influence. Its value ranges from 0 to 1, and it adjusts the weight of the historical data relative to the current data Ibrahim et al. [15]. A value of $\alpha_0 = 0$ implies that the historical data have no influence on the analysis, effectively ignoring the prior data and using as a prior for the current analysis the same prior as in the previous analysis $\pi_0(\boldsymbol{\theta})$. Alternatively, a value of $\alpha_0 = 1$ treats the historical data with full weight, as if they were part of the current data set. Intermediate values between 0 and 1 allow for a weighted incorporation of historical information, with higher values assigning more weight to the historical data.

As α_0 is external to the problem itself, there must be some principled method by which it is selected. We address this problem in Sect. 6.2 on prior selection.

4.2 Gaussian Process Regression

To smooth the collection of estimated coefficients, Gaussian process regression (GPR) was used. GPR is based on Gaussian processes (GPs) [25], a family of continuous-time, continuous-state stochastic processes that can be used to model distributions over function spaces. More specifically, a Gaussian process is a collection of random variables $\{X_t, t \in \mathbb{R}\}$, any finite number of which have a joint multivariate normal (Gaussian) distribution [25].

The family of normal distributions is parameterized by the distributions' means $\mu \in \mathbb{R}$ and variances $\sigma^2 > 0$. The family of multivariate normal distributions is parameterized by the distributions' multivariate means $\boldsymbol{\mu} \in \mathbb{R}^p$ and covariance matrices $\boldsymbol{\Sigma} \in \mathbb{S}_+^p$, where $\mathbb{S}_+^p$ is the set of $p \times p$ real symmetric positive definite matrices. Generalizing further, the family of Gaussian processes is parameterized by a mean function $\mu(t)$ and a covariance function $\sigma(s, t)$, a function such that when applied to any pair of a finite collection of values $t_1, \ldots, t_n \in \mathbb{R}$ results in a real symmetric positive definite matrix. There are several common functions used for the covariance function σ, we revisit this in a moment.

Gaussian processes share with multivariate normal distributions the key property that the conditional distribution of a finite number of the random variables of the process $\boldsymbol{X}_1 = [X_{t_1}, \ldots, X_{t_n}]$ given another finite collection of random variables of the process $\boldsymbol{X}_2 = [X_{s_1}, \ldots, X_{s_{n'}}]$ is multivariate normal, where the parameters of the multivariate normal depend in the on the observed quantities and the parameters of the process in standard way, namely if $\boldsymbol{\mu} = [\boldsymbol{\mu}_1 \ \boldsymbol{\mu}_2]$ and $\boldsymbol{\Sigma} = [\boldsymbol{\Sigma}_{11}, \boldsymbol{\Sigma}_{12}; \boldsymbol{\Sigma}_{21}, \boldsymbol{\Sigma}_{22}]$ are the mean and covariance functions evaluated at the values $t_1, \ldots, t_n, s_1, \ldots, s_{n'}$, then

$$[\boldsymbol{X}_1 | \boldsymbol{X}_2 = \boldsymbol{x}_2] \sim \mathcal{N}_n \left(\boldsymbol{\mu}_1 + \boldsymbol{\Sigma}_{12} \boldsymbol{\Sigma}_{22}^{-1} (\boldsymbol{x}_2 - \boldsymbol{\mu}_2), \boldsymbol{\Sigma}_{11} - \boldsymbol{\Sigma}_{12} \boldsymbol{\Sigma}_{22}^{-1} \boldsymbol{\Sigma}_{21} \right).$$

Gaussian processes as described interpolate the data. Gaussian process regression (GPR), by contrast, introduces an additional variance term that loosens this restriction to provide a flexible non-parametric Bayesian approach to curve fitting. In particular, GPR involves the covariance function $K(s,t)$ defined as a modification of $\sigma(s,t)$:

$$K(s,t|\gamma,\sigma^2) = \sigma(s,t|\gamma) + \delta_{s,t}\sigma^2, \tag{4}$$

where $\delta_{s,t}$ is the Kronecker delta function equal to 1 when $s = t$ and 0 otherwise, and γ is any parameters that the core covariance function, the kernel, employs. Many choices are common for the kernel covariance function $\sigma(s,t|\gamma)$, here we adopt the squared exponential (RBF) kernel with the form [21, 28]

$$K(s,t|\gamma,\sigma^2) = K(s,t|\alpha,\rho,\sigma^2) = \alpha^2 \exp\left(-\frac{1}{2\rho^2}(s-t)^2\right) + \delta_{i,j}\sigma^2, \tag{5}$$

where α governs the overall amplitude of the covariance function. Higher values for α means that the function is more flexible and can fit the data with higher variance, while lower values results in a smoother, less varying function. ρ determines the distance over which the covariance between data points attenuates. A smaller value of ρ implies that data points close to each other will have high covariance, leading to more local variations in the function. Conversely, a larger value of ρ results in a smoother function, as the covariance decreases more rapidly with distance. σ^2 represents the noise or uncertainty present in the observations. It adds a diagonal term to the covariance matrix, accounting for the variability in the observed data points and ensuring that each data point has a noise term associated with it.

We used exponentiated quadratic function for the covariance function because of its smoothness, fewer number of hyperparameters and easier implementation in Stan, our main Bayesian computational engine. The kernel has only a couple of hyperparameters, which simplifies the model.

5   Model Comparison

In this section, we outline our approach for comparing the newly proposed Bayesian model to the model presented by Asif and McHale [2]. We compared the two models in terms of accuracy and Brier scores based on 10-fold cross-validation. Explicitly, we partitioned the data into training (90%) and testing (10%) sets, fit both the models with the training data and made predictions with the testing data. We repeated this process 10 times, each time with different set of training and testing division of data. We then proceeded to compare the accuracy and Brier scores of the predictions.

5.1   Brier Scores

The Brier score is a statistical measure of accuracy of probabilistic predictions particularly useful in evaluating the performance of prediction models [5]. The essence of the Brier score is to quantify the accuracy of probabilistic forecasts by comparing the predicted probabilities against the actual outcomes. It does so by calculating the mean squared difference between the predicted probability assigned to an event occurring and the actual outcome (with the outcome being 1 if the event occurs and 0 if it does not). This differentially penalizes incorrect predictions based on their uncertainty, with very confident incorrect predictions penalized more.

The Brier score is defined to be:

$$\text{Brierscore} = \frac{1}{N} \sum_{i=1}^{N} (P_i - O_i)^2,$$

where N is the total number of instances or observations, P_i is the predicted probability for the i-th instance, and O_i is the observed outcome for the i-th instance, typically represented as either 0 or 1, a binary outcome.

Brier scores are simple and easy to interpret. For each prediction, you calculate the square of the difference between the predicted probability and the actual outcome, and then average those squared differences over all predictions. The squaring ensures that the score is always positive and penalizes larger errors more severely. The Brier score ranges from 0 to 1, with 0 indicating a perfect match between predicted probabilities and observed outcomes (ideal calibration), and 1 representing the worst possible score (no calibration). We used 10-fold cross-validated Brier score and accuracy to compare our models in Sect. 6.

6   Application

This section involves discussion about the application of the models discussed in Sect. 4 and its comparison with the on in Sect. 3. The data that we applied both the models to comes from the Indian Premier League (IPL) from season 2010 to 2020 as discussed in Sect. 2.

6.1   Variable Selection

As explained in Sect. 3.3, the process began with identifying the optimal subset of candidate covariates for inclusion in the model through delete-d cross validation (CVd) using random subsets prior to fitting the sequence of K independent logistic models for an innings. Subsequently, covariates that appeared at least once in the best logistic model in each innings were selected for further analysis. Table 3 presents the number of times that each covariate appeared in the best logistic model in each innings. Interestingly, the factor of playing at home was found to be significant for teams chasing a target, but not for those batting first. Conversely, playing away held importance for teams that batted first. The covariates representing toss, form difference (FD), and rating difference (RD) did not show significance at any stage of the games. The variable representing wickets resource lost (WRL) emerged as the most critical factor in determining win probability for both teams, irrespective of batting order. Moreover, runs per over (RPO) and required runs per over (RRPO) were significant factors for teams batting first and second, respectively. We selected away, WRL and RPO for first innings and home, WRL and RRPO for second innings.

Table 3. Number of times (or balls) that each covariate appeared in the best logistic model ineach innings.

Innings	Home	Away	Toss	RD	FD	WRL	RPO	RRPO
First	0	87	0	0	0	89	95	NA
Second	24	0	0	0	0	117	NA	120

6.2 Priors

In this section, we talk about the priors that were selected for our models. In the case of historical data, we did not have any information so we opted to use diffuse normal priors $\mathcal{N}(0, \sigma = 10)$ on both intercept and slope terms. A normal prior with mean at 0 and standard deviation of 10 for logistic regression parameters starts from a neutral position, treating positive and negative effects as equally likely, and its wide standard deviation is close to flat.

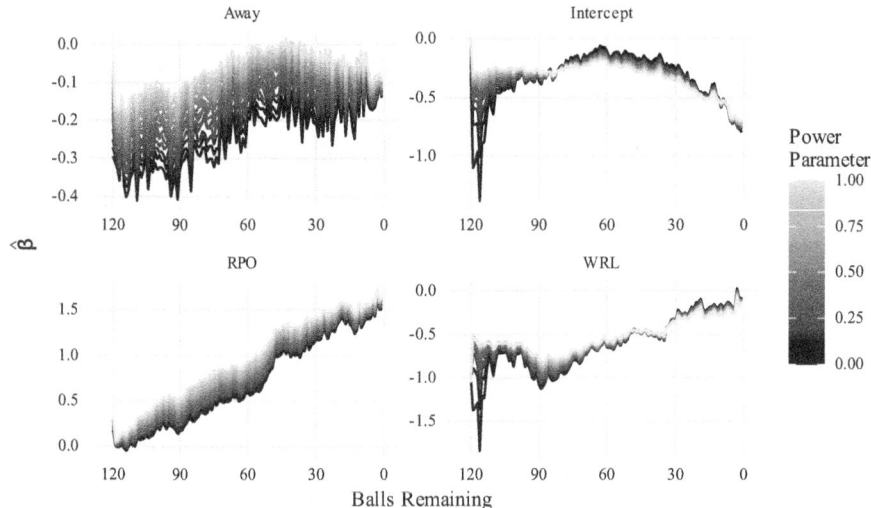

Fig. 3. Mean estimates using different power parameters for first innings.

As described above, power priors control the influence of the historical data on the analysis of the current data, acting as a weighting factor. To assess the effect of having different values of power parameter α_0 in the regression coefficients before choosing a value, we ran the analysis for many different values of the power parameter and plotted the resulting curves $\widehat{\beta}_m(k)$. In Fig. 3, we can see the mean estimates for different covariates using different power parameter values at different stages of the first innings. We can clearly see that there is significant amount of variation in the estimates when using different values of power parameter, indicating a moderate level of borrowing from past data. This phenomenon was consistent with different sets of data and not just this specific simulation where we used all the data from 2008 to 2018 for modelling. The same variations were observed for the second innings as well. To avoid having to specify a specific value, we opted to put a prior on the power parameter since selecting power parameter for all 120 models for each innings would be challenging. For this, we used uniform Beta $(1, 1)$ distribution.

When addressing the covariance function in the context of (4), we required priors for three parameters: ρ, α, and σ^2. For α, which corresponds to the marginal standard deviation in a Gaussian process, we selected the half standard normal prior. This prior places meaningful probability mass near zero, supporting models where the Gaussian process might contribute minimally or even zero to the conditional mean of the output, thus allowing for flexibility in how much of the variation the process explains without dominating the response. We also used half standard normal distribution for σ in order to ensure that σ stays positive while allowing for flexibility in noise noise or uncertainty present in the observations. As for ρ, we used an inverse gamma distribution with shape parameter 5 and scale parameter 5 in order to ensures a moderate restriction, preventing overly small values (which cause overfitting) while still allowing for sufficient variability to capture broader trends in the data..

6.3   Comparison

Our analysis involved evaluating the two models through 10-fold cross-validation techniques, focusing on Accuracy and Brier score metrics as outlined in Sect. 5. Before this in-depth comparison, we initiated our examination by predicting the outcome of an individual game using both models. This process began by randomly selecting a match for analysis, which in this case was the contest between Kolkata Knight Riders and Sunrisers Hyderabad on April 14, 2017. Kolkata, batting first, set a target of 171 runs at the expense of 8 wickets. In response, Hyderabad managed to score 155 runs, losing only 2 wickets over the course of their 20 overs. Figure 4 showcases the win probabilities predicted by both models at various stages of the match. It is noteworthy that the team batting first clinched the victory in this instance. A detailed review of the graphical data reveals that our Bayesian model consistently provided modestly higher win probabilities for the winning side, especially during the early phases of the first innings, when compared to the predictions made by Asif and McHale's approach. Although the disparity between the models was modest, our Bayesian model demonstrated a marginally superior capability in forecasting the game's outcome.

As the match unfolded into the second innings and concluded with the defeat of the team batting second, our attention was drawn to the models' predictions of decreasing win probabilities for the trailing team. Both models exhibited similar predictive behaviors towards the climax of the second innings. This can be expected as the games mostly are decided at this time unless there is a major

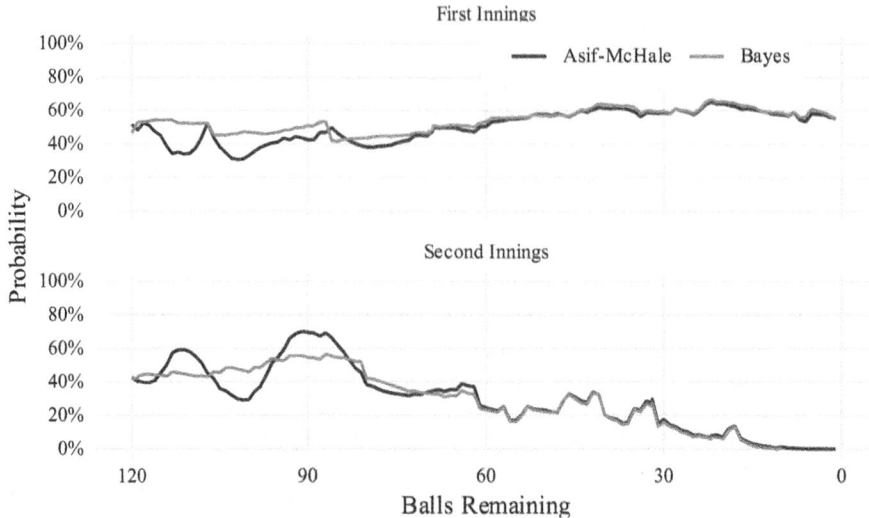

Fig. 4. In-game win probability for a single game (Kolkota vs Hyderbad)

upset. However, at the beginning, the Bayesian model seems to be more stable than the model by Asif and McHale [2]. It is not quite clear why, but prior data may be serving to create more stable estimates for this point in the game.

Our Bayesian model performed slightly better in the single game picked at random as shown by Fig. 4. However, this of course does not speak to the general behavior or relative performance of the models; for this we need to look at all the matches. Figure 5 has plots the result for 10-fold cross validated raw accuracy for both the model in the innings.

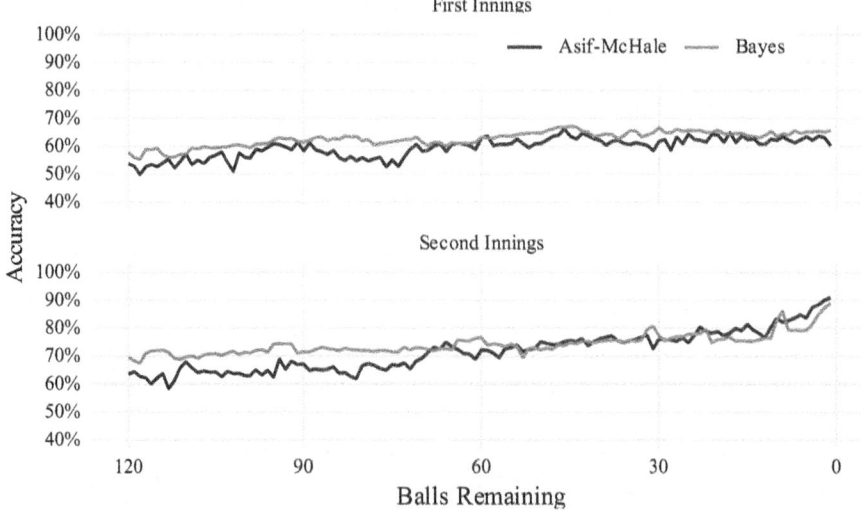

Fig. 5. 10-fold Cross Validated Accuracy

The chart clearly demonstrates that our model consistently though modestly outperforms the alternative model in terms of accuracy across the majority of game stages. It is worth noting that, towards the end stages, both models exhibited similar levels of accuracy. Nevertheless, throughout the game, the Bayesian model consistently maintained a better level of accuracy. This is generally to be expected, since the models are largely the same.

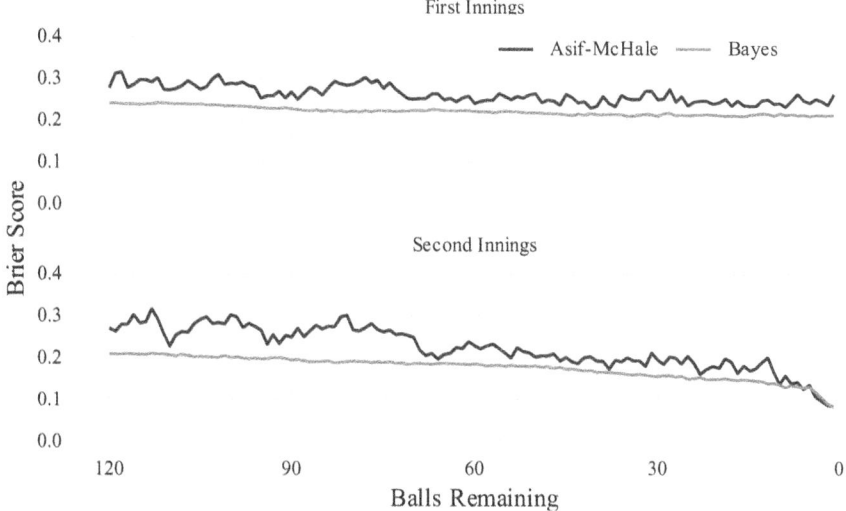

Fig. 6. 10-fold Cross Validated Brier Scores

Exploring further into the nuances of model assessment, Fig. 6 presents a look at the Brier score through 10-fold cross-validation analysis. A reduced Brier score indicates a more accurate model. The figure clearly shows that the Bayesian model consistently achieves a lower Brier score across every phase of the game. This trend suggests that the probabilistic forecasts of the Bayesian model are more closely matched with the real-world results, highlighting its improved predictive power over the approach by Asif and McHale [2].

7   Conclusion

As sports analytics gains more traction in academic circles, we can expect more sophisticated models to enter into sports. Other advanced methods, including machine learning and neural networks, could also enhance the predictive power of the models. However, we did not employ these techniques because the goal of this research was to build upon the existing work by Asif and McHale. Additionally, our data availability was limited, as most machine learning models require large datasets. Furthermore, we aimed to develop a model that would not only be predictive but also provide a more transparent and interpretable understanding of the relationships between the variables, allowing for clearer insights and more actionable conclusions.

In this article we have considered Bayesian methods using power priors and Gaussian processes for in-game win prediction of cricket matches. The results of our analysis suggest that Bayesian methods can provide a flexible framework with which to incorporate past data into a cricket analysis. The model albeit slightly was able to improve the accuracy of the predictions and lead us to believe that it can be a substantial tool in the field of sports analytics.

There are a number of things that can be explored going forward. We could possibily look the game from bowling point of view as well. Using individual player attributes instead of team could be another place to examine. Incorporating the "12th man" rule that has been included in the latest renditions of IPL which allows a team to swap a player at any moment, would be another interesting problem. From modelling point of view, we could investigate use of different covariance matrix function on the Gaussian processes, different priors on the covariance matrices or different prior elicitation process altogether. The versatility of Bayesian methods confirms their potential for improving the field of cricket and sports anlytics.

References

1. Akhtar, S., Scarf, P.: Forecasting test cricket match outcomes in play. Int. J. Forecast. **28**(3), 632–643 (2012). ISSN 01692070, https://doi.org/10.1016/j.ijforecast.2011.08.005, https://linkinghub.elsevier.com/retrieve/pii/S0169207011001622

2. Asif, M., McHale, I.G.: In-play forecasting of win probability in one-day international cricket: a dynamic logistic regression model. Int. J. Forecast. **32**(1), 34–43 (2016). ISSN 01692070, https://doi.org/10.1016/j.ijforecast.2015.02.005, https://linkinghub.elsevier.com/retrieve/pii/S0169207015000618

3. Bailey, M., Clarke, S.R.: Predicting the match outcome in one day international cricket matches while the game is in progress (2006)

4. Boys, R.J., Philipson, P.M.: On the ranking of test match batsmen, June 2018. https://doi.org/10.48550/arXiv.1806.05496, http://arxiv.org/abs/1806.05496, arXiv:1806.05496 [stat]

5. Brier, G.W.: Verification of forecasts expressed in terms of probability. Monthly Weather Rev. **78**(1), 1–3 (1950). ISSN 1520-0493, 0027-0644, https://doi.org/10.1175/1520-0493(1950)078<0001:VOFEIT>2.0.CO;2, publisher: American Meteorological Society Section: Monthly Weather Review

6. Brooks, R.D., Faff, R.W., Sokulsky, D.: An ordered response model of test cricket performance. Appl. Econ. **34**(18), 2353–2365 (2002). ISSN 0003-6846, 1466-4283, https://doi.org/10.1080/00036840210148085, http://www.tandfonline.com/doi/abs/10.1080/00036840210148085

7. Brown, L.D.: In-season prediction of batting averages: a field test of empirical Bayes and Bayes methodologies. Ann. Appl. Stat. **2**(1), 113–152 (2008). ISSN 1932-6157, 1941-7330, https://doi.org/10.1214/07-AOAS138, https://projecteuclid.org/journals/annals-of-applied-statistics/volume-2/issue-1/In-season-prediction-of-batting-averages--A-field-test/10.1214/07-AOAS138.full, publisher: Institute of Mathematical Statistics

8. cricsheet: Available match data downloads - Cricsheet (2023). https://cricsheet.org/downloads/

9. Duckworth, F.C., Lewis, A.J.: A fair method for resetting the target in interrupted one-day cricket matches. J. Oper. Res. Soc. **49**(3), 220–227 (1998). ISSN 0160-5682, 1476-9360, https://doi.org/10.1057/palgrave.jors.2600524, https://www.tandfonline.com/doi/full/10.1057/palgrave.jors.2600524

10. Duckworth, F.C., Lewis, A.J.: A successful operational research intervention in one-day cricket. J. Oper. Res. Soc. **55**(7), 749–759 (2004). ISSN 0160-5682, 1476-9360, https://doi.org/10.1057/palgrave.jors.2601717, https://www.tandfonline.com/doi/full/10.1057/palgrave.jors.2601717

11. Efron, B., Morris, C.: Combining possibly related estimation problems. J. Royal Stat. Soc. Ser. B (Methodological) **35**(3), 379–402 (1973). ISSN 00359246, https://doi.org/10.1111/j.2517-6161.1973.tb00968.x, https://onlinelibrary.wiley.com/doi/10.1111/j.2517-6161.1973.tb00968.x

12. Elo, A.: The Rating of Chess Players, Past & Present. Arco Publishing, Inc., 2 edn. (1978)

13. Elo, A.E.: The proposed USCF rating system: Its development, theory, and applications. Chess Life **22**(8), 242–247 (1967)

14. Heinzen, E.: ELO: ranking teams by elo rating and comparable methods (2023). https://CRAN.R-project.org/package=elo, r package version 3.0.2

15. Ibrahim, J.G., Chen, M., Gwon, Y., Chen, F.: The power prior: theory and applications. Stat. Med. **34**(28), 3724–3749 (2015). ISSN 0277-6715, 1097-0258, https://doi.org/10.1002/sim.6728, https://onlinelibrary.wiley.com/doi/10.1002/sim.6728

16. Jensen, S.T., McShane, B.B., Wyner, A.J.: Hierarchical bayesian modeling of hitting performance in baseball. Bayesian Anal. **4**(4) (2009). ISSN 1936-0975, https://doi.org/10.1214/09-BA424, https://projecteuclid.org/journals/bayesian-analysis/volume-4/issue-4/Hierarchical-Bayesian-modeling-of-hitting-performance-in-baseball/10.1214/09-BA424.full

17. Jiang, W., Zhang, C.H.: Empirical Bayes in-season prediction of baseball batting averages. In: Institute of Mathematical Statistics Collections, pp. 263–273, Institute of Mathematical Statistics, Beachwood, Ohio, USA (2010). ISBN 978-0-940600-79-9, https://doi.org/10.1214/10-IMSCOLL618, http://projecteuclid.org/euclid.imsc/1288099025

18. Kaggle: IPL Complete Dataset (2008-2020) (2023). https://www.kaggle.com/datasets/patrickb1912/ipl-complete-dataset-20082020

19. Karlis, D., Ntzoufras, I.: Bayesian modelling of football outcomes: using the Skellam's distribution for the goal difference. IMA J. Manage. Math. **20**(2), 133–145 (2008). ISSN 1471-678X, 1471-6798, https://doi.org/10.1093/imaman/dpn026, https://academic.oup.com/imaman/article-lookup/doi/10.1093/imaman/dpn026

20. Morley, B., Thomas, D.: An investigation of home advantage and other factors affecting outcomes in English one-day cricket matches. J. Sports Sci. **23**(3), 261–268 (2005). ISSN 0264-0414, 1466-447X, https://doi.org/10.1080/02640410410001730133, http://www.tandfonline.com/doi/abs/10.1080/02640410410001730133

21. Murphy, K.P.: Machine Learning: a Probabilistic Perspective. MIT Press, Cambridge (2012)

22. Neal, D., Tan, J., Hao, F., Wu, S.S.: Simply better: using regression models to estimate major league batting averages. J. Quant. Anal. Sports **6**(3) (2010). ISSN 1559-0410, https://doi.org/10.2202/1559-0410.1229, https://www.degruyter.com/document/doi/10.2202/1559-0410.1229/html

23. Nishanth, S.: Ranking the Indian Premier League (IPL) teams using the ELO rating system, May 2020. https://sathvik-nishanth.medium.com/ranking-the-indian-premier-league-ipl-teams-using-the-elo-rating-system-3bbfc2f14605

24. R Core Team: R: A Language and Environment for Statistical Computing. R Foundation for Statistical Computing, Vienna, Austria (2023). https://www.R-project.org/

25. Rasmussen, C.E., Williams, C.K.I.: Gaussian Processes for Machine Learning. The MIT Press (2005), ISBN 978-0-262-25683-4, https://doi.org/10.7551/mitpress/3206.001.0001, https://direct.mit.edu/books/book/2320/gaussian-processes-for-machine-learning

26. Santos-Fernandez, E., Wu, P., Mengersen, K.L.: Bayesian statistics meets sports: a comprehensive review. J. Quant. Anal. Sports **15**(4), 289–312 (2019). ISSN 1559-0410, 2194-6388, https://doi.org/10.1515/jqas-2018-0106, https://www.degruyter.com/document/doi/10.1515/jqas-2018-0106/html

27. Scarf, P., Shi, X., Akhtar, S.: On the distribution of runs scored and batting strategy in test cricket. J. R. Stat. Soc. Ser. A **174**, 471–497 (2011)

28. Stan Development Team: Stan User's Guide Version 2.34. Stan Developers (2024). https://mc-stan.org/docs/stan-users-guide/

29. Stevenson, O.G., Brewer, B.J.: Bayesian survival analysis of batsmen in test cricket, February 2017. https://doi.org/10.48550/arXiv.1609.04078, http://arxiv.org/abs/1609.04078, arXiv:1609.04078 [stat]

30. Suzuki, A.K., Salasar, L.E.B., Leite, J.G., Louzada-Neto, F.: A Bayesian approach for predicting match outcomes: the 2006 (Association) Football World Cup. J. Oper. Res. Soc. **61**(10), 1530–1539 (2010). ISSN 0160-5682, 1476-9360, https://doi.org/10.1057/jors.2009.127, https://www.tandfonline.com/doi/full/10.1057/jors.2009.127

Finite-Time Outer Average Synchronization Between Two Coupled Heterogeneous Complex Dynamical Networks and Its Application in Secure Communication

Lihong Yan(✉) ⓘ

College of Mathematics and Statistics, Xianyang Normal University, Xianyang 712000, Shannxi, People's Republic of China
092212@163.com

Abstract. The finite-time outer average synchronization control problem for a class of heterogeneous complex dynamical networks with unknown parameters and its application in secure communication is addressed in this paper. Basing on finite-time stabilization theorem of differential equations system, the decentralized adaptive feedback control strategy and properties of inequalities for vectors, some useful finite-time outer average synchronization criteria have been obtained. Finally, an illustrative example is used to demonstrate the efficiency and effectiveness of the developed method.

Keywords: Complex Dynamical Network · Outer Average Synchronization · Finite-time · Adaptive Control · Secure Communication

1 Introduction

As large-scale dynamical systems, complex dynamical networks have being one of the research hot spots in control field, researchers mainly concerned about model building, properties analysis, dynamical evolution and synchronization control of complex dynamical networks [1, 2]. Among the problems concerned, synchronization is one of the most interesting topics and has been extensively investigated over few past years. The synchronization mechanism investigated from the point of view of nonlinear dynamics help researchers to found more avenues to enhance the synchronous ability of networks [3–5], and reduce its negative factors in the field of physics, biology, laser systems, communication systems and so on.

One classical synchronization strategy for complex dynamical networks is injecting external controller designed into the network, such as impulse control, event-triggered strategy, intermittent control and sample-data control etc. While feedback control, adaptive control, fuzzy control, optimal control and sliding mode control are continuous control technique, these traditional control methods provide us a large mount of synchronization conclusions and help us to find more about the complex dynamical network deeply.

H. Han (Ed.): SDSC 2024, CCIS 2158, pp. 169–182, 2024.
https://doi.org/10.1007/978-3-031-67871-4_12

Guo and Li [6] proposed a new adaptive learning control approach for the synchronization of complex dynamical networks with unknown time-varying parameters. By using the reparameterization technique, the adaptive learning laws of periodically time-varying and constant parameters and an adaptive control strategy are designed to ensure the asymptotic convergence of the synchronization error. In 2015, Tan and Tian [7] proposed the finite time stabilization and synchronization conclusions for drive-response complex networks with different dimensions based on finite time stability theory. In [8], Zhang et al. achieved matrix projection synchronization of a complex network model with unknown node parameters and disturbances, as well as different node dimensions, using adaptive control methods. Hao and Li [9] investigated the stochastic synchronization problem for complex dynamical networks with unknown periodic time-varying couplings and with stochastic noise perturbations.

Among all studies mentioned above, the synchronization of complex dynamical network is asymptotic in square mean or probability sense. However, in numerous actual applications, it is always expected that the synchronization can be achieved as quickly as possible, so the finite-time control technique is adopted. Based on the robustness and anti-interference characteristics of finite-time control technology, Yang and Cao [10] studied the finite-time stochastic synchronization of complex networks with random noise interference. In [11], Sun et al. analyzed the finite-time stochastic outer synchronization between two nonidentical dynamical networks with stochastic disturbances. Zhao et al. [12] investigated stochastic mean square modified function projective synchronization of uncertain complex network with multi-links. Yan and Li [13] proposed the definition of outer average synchronization (OAS in brief) between two coupled networks, based on LaSalle's invariance principle, unknown parameters both of the drive and response networks are identified via adaptive control strategy.

It is worth noting that, there is little literature above considering both the finite-time outer average synchronization and unknown parameter estimations of complex dynamical networks. So in this paper, we will consider the finite-time outer average synchronization (FTOAS in brief) for drive-response complex dynamical networks with unknown parameters by using an adaptive feedback controllers with the help of the advantages of feedback control, adaptive control and learning control. Also an image encryption and decryption algorithm based on finite-time outer average synchronization of complex dynamical networks in secure communication is presented finally.

2 Problem Statement and Preliminaries

In the paper, we will consider the following coupled CDNs, the drive network is,

$$\dot{x}_i(t) = f_i(x_i(t)) + F_i(x_i(t))\alpha_i + c_1 \sum_{j=1}^{N} a_{ij}\Gamma_1 x_j(t), \ i = 1, 2, \cdots, N \qquad (1)$$

where $x_i(t) = (x_{i1}(t), x_{i2}(t), \cdots, x_{in}(t))^T \in \mathbb{R}^n$ is the state variable of the i−th node. $f_i(x_i(t)) : \mathbb{R}^n \to \mathbb{R}^n, i = 1, 2, \cdots, N$ is a smooth nonlinear function without parameters, $F_i : \mathbb{R}^n \to \mathbb{R}^{n \times m}$ is function matrices, $\alpha_i, i = 1, 2, \cdots, N$ are unknown parameters, $f_i(x_i(t)) + F_i(x_i(t))\alpha_i$ describes the local dynamics of each node of the drive network. c_1 is outer coupling strength, $\Gamma_1 = diag(\gamma_1^1, \gamma_2^1, \cdots, \gamma_n^1)$ is inner coupling matrix.

And the response network with different numbers of nodes is,

$$\dot{y}_i(t) = g_i(y_i(t)) + G_i(y_i(t))\beta_i + c_2 \sum_{j=1}^{M} b_{ij}\Gamma_2 y_j(t) + u_i(t), \quad i = 1, 2, \cdots, M \quad (2)$$

where $y_i(t) = (y_{i1}(t), y_{i2}(t), .., y_{in}(t))^T \in \mathbb{R}^n$ is the state variable of node i, $g_i(y_i(t))$: $\mathbb{R}^n \to \mathbb{R}^n$, $i = 1, 2, \cdots, M$ is a smooth non-linear function, $G_i : \mathbb{R}^n \to \mathbb{R}^{n \times m}$ are function matrices. $\beta_i, i = 1, 2, \cdots, M$ are unknown bounded parameters, c_2 outer coupling strength, $\Gamma_2 = diag(\gamma_1^2, \gamma_2^2, \cdots, \gamma_n^2)$ is inner coupling matrix, $u_i(t)$, $i = 1, 2, \cdots, M$ are outer controllers for each node.

In addition, suppose the outer-coupling configurations of drive and response networks (1) and (2) are both bidirectional [14], i.e. if there is a link from node i to node j, $a_{ji} > 0$. While if there is a link from node j to node i, $a_{ij} > 0$. The columns' sums of outer coupling matrices are all equals to zero. That is,

$$\begin{cases} \sum_{i=1,i\neq j}^{N} a_{ij} = -a_{jj}, j = 1, 2, \cdots, N \\ \sum_{i=1,i\neq j}^{M} b_{ij} = -b_{jj}, j = 1, 2, \cdots, M \end{cases} \quad (3)$$

In the paper, we will design proper controllers $u_i(t), i = 1, 2, \cdots, M$ to realize FTOAS between the drive and response networks (1) and (2). For convenience, the following definitions and lemma need to be introduced firstly.

The average state orbits of the drive and response systems (1) and (2) separately are defined as follows,

$$\bar{x}(t) = \frac{1}{N} \sum_{i=1}^{N} x_i(t), \ \bar{y}(t) = \frac{1}{M} \sum_{i=1}^{M} y_i(t) \quad (4)$$

The outer average error is defined as following,

$$\bar{e}(t) = \bar{y}(t) - \bar{x}(t) = \frac{1}{M} \sum_{i=1}^{M} y_i(t) - \frac{1}{N} \sum_{i=1}^{N} x_i(t) \quad (5)$$

Definition 1. It is said that the two coupled networks (1) and (2) realize FTOAS means that, if there exists a constant $t^* > 0$, such that

$$\begin{cases} \lim_{t \to t^*} \|\bar{y}(t) - \bar{x}(t)\| = 0, \\ \|\bar{y}(t) - \bar{x}(t)\| \equiv 0, \quad t \geq t^* \end{cases} \quad (6)$$

Moreover, The uncertain system parameter vectors $\hat{\alpha}_i, \hat{\beta}_i$ are identified in a finite time if there exists a constant t^*, such that

$$\lim_{t \to t^*} \|\hat{\alpha}_i(t) - \alpha_i(t)\| = 0, \quad \lim_{t \to t^*} \|\hat{\beta}_i(t) - \beta_i(t)\| = 0 \quad (7)$$

The identification parameters errors $\tilde{\alpha}_i(t)$, $\tilde{\beta}_i(t)$ are defined as: $\tilde{\alpha}_i(t) = \hat{\alpha}_i(t) - \alpha_i(t)$, $i = 1, 2, \cdots, N$, $\tilde{\beta}_i(t) = \hat{\beta}_i(t) - \beta_i(t)$, $i = 1, 2, \cdots, M$. where $\hat{\alpha}_i(t)$, $\hat{\beta}_i(t)$ are the estimations of unknown parameters.,

So the FTOAS error dynamical function between the networks (1) and (2) is as following,

$$\dot{\bar{e}}(t) = \frac{1}{M} \sum_{i=1}^{M} \dot{y}_i(t) - \frac{1}{N} \sum_{i=1}^{N} \dot{x}_i(t)$$

$$= \bar{g}(t) - \bar{f}(t) + \frac{1}{M} \sum_{i=1}^{M} G_i(y_i(t))\beta_i - \frac{1}{N} \sum_{i=1}^{N} F_i(x_i(t))\alpha_i + \frac{1}{M} \sum_{i=1}^{M} u_i(t) \quad (8)$$

where $\bar{f}(t) = \frac{1}{N} \sum_{i=1}^{M} G_i(y_i(t))\beta_i$, $\bar{g}(t) = \frac{1}{M} \sum_{i=1}^{M} g_i(y_i(t))$. It is easy to see that FTOAS between the drive-response networks (1) and (2) is equal to the error system (8) is stable at the original. For further discussion, several lemmas are introduced.

Lemma 1 [15] . Assume that a continuous, positive-definite function $V(t)$ satisfies the following differential inequality,

$$\dot{V}(t) \leq -\alpha V^{\eta}(t) + \beta V(t), \ \forall t \geq 0, \ V^{1-\eta}(0) \leq \frac{\alpha}{\beta} \quad (9)$$

where $0 < \eta < 1, \alpha > 0, \beta > 0$, then for $0 \leq t \leq t^*$, the following inequality holds,

$$e^{-\beta(1-\eta)t} V^{1-\eta}(t) \leq V^{1-\eta}(0) - \frac{\alpha}{\beta} + \frac{\alpha}{\beta} e^{-(1-\eta)\beta t} \quad (10)$$

the settling time t^* satisfy

$$t^* \leq \frac{\ln\left(1 - \frac{\beta}{\alpha} V^{1-\eta}(0)\right)}{\beta(\eta - 1)} \quad (11)$$

Lemma 2 [16] . Assume that $\rho_i, i = 1, 2, \cdots, n$ are positive numbers, σ is positive satisfying $0 < \sigma < 1$, then the following inequality holds,

$$\sum_{i=1}^{n} \|\rho_i\|^{\sigma} \leq \left(\sum_{i=1}^{n} \|\rho_i\|^2\right)^{\frac{\sigma}{2}}$$

Assumption 1. There exist non-negative constants B_{α}, B_{β}, such that,

$$\begin{cases} \alpha_i(t) \leq B_{\alpha} \\ \beta_i(t) \leq B_{\beta} \end{cases} \quad (12)$$

3 Finite-Time Outer Average Synchronization Between Two Coupled Networks

In this part, we consider FTOAS between coupled complex dynamical network (1) and (2) with adaptive control approach based on the finite-time synchronization theorem.

Theorem 1. Suppose Lemma 1 holds, the drive and response dynamical networks (1) and (2) can realize FTOAS by the following nonlinear adaptive feedback controllers (13) and the uncertain system parameter vectors can be identified simultaneously by adopting the updated laws (14).

$$
\begin{cases}
u_{i1}(t) = \bar{f}(t) - \bar{g}(t) + \dfrac{1}{N}\sum_{i=1}^{N} F_i(x_i(t))\hat{\alpha}_i - \dfrac{1}{M}\sum_{i=1}^{M} G_i(y_i(t))\hat{\beta}_i - \dfrac{\sigma}{\sqrt{2^{1+\eta}}}\,\mathrm{sign}(\bar{e}(t))\,|\bar{e}(t)|^{\eta} \\[2mm]
\qquad - \dfrac{\sigma}{\sqrt{(2N)^{1+\eta}}}\sum_{i=1}^{N} \dfrac{\left(\|\hat{\alpha}_i\| + B_\alpha\right)^2}{l_i}\,\dfrac{\bar{e}(t)}{\|\bar{e}(t)\|^2} - \dfrac{\sigma}{\sqrt{(2M)^{1+\eta}}}\sum_{i=1}^{N} \dfrac{\left(\|\hat{\beta}_i\| + B_\beta\right)^2}{m_i}\,\dfrac{\bar{e}(t)}{\|\bar{e}(t)\|^2} \\[2mm]
u_{i2}(t) = -k_i(t)\bar{e}(t) \qquad\qquad\qquad\qquad\qquad\qquad\qquad\qquad\quad \text{if } \|\bar{e}(t)\| \neq 0 \\[1mm]
\begin{cases} u_{i1}(t) = 0 \\ u_{i2}(t) = 0 \end{cases} \qquad\qquad\qquad\qquad\qquad\qquad\qquad\qquad\quad \text{if } \|\bar{e}(t)\| = 0 \\[3mm]
\dot{k}_i(t) = -\dfrac{\sigma}{\sqrt{2^{1+\eta}}}\, r_i^{\frac{1-\eta}{2}}\,|k_i|^{\eta}\,\mathrm{sign}(k_i) + \dfrac{r_i}{M}\left(1 + \dfrac{\mu_i}{k_i}\right)\bar{e}^T(t)\bar{e}(t)
\end{cases}
\tag{13}
$$

where σ, μ_i, k_i, r_i are positive, the real number η satisfies $0 < \eta < 1$, and the parameter adaptive update law are,

$$
\begin{cases}
\dot{\hat{\alpha}}_i(t) = -l_i F_i^T(x_i(t))\bar{e}(t), & i = 1, 2, \cdots, N \\[1mm]
\dot{\hat{\beta}}_i(t) = m_i G_i^T(y_i(t))\bar{e}(t), & i = 1, 2, \cdots, M
\end{cases}
\tag{14}
$$

where $u_{i1}(t)$ and $u_{i2}(t)$, $i = 1, 2, \cdots, M$ are the non-linear feedback controllers and the linear adaptive feedback controllers, respectively and l_i, m_i are any positive constants. Then the synchronization and estimation of drive-response network (1) and (2) can be achieved in the finite time

$$
t^* \leq \frac{2\ln\left(1 - \frac{\mu}{\sigma}V^{\frac{1-\eta}{2}}(0)\right)}{\mu(\eta - 1)}
$$

Where $V(0) = \frac{1}{2}\bar{e}^T(0)\bar{e}(0) + \frac{1}{2N}\sum_{i=1}^{N}\frac{1}{l_i}\tilde{\alpha}_i^2(0) + \frac{1}{2M}\sum_{i=1}^{N}\frac{1}{m_i}\tilde{\beta}_i^2(0) + \frac{1}{2}\sum_{i=1}^{N}\frac{1}{r_i}k_i^2(0)$, $\mu = \frac{2}{M}\sum_{i=1}^{M}\mu_i$, $\tilde{\alpha}_i(0)$, $\tilde{\beta}_i(0)$, $k_i(0)$ are initial conditions.

In fact, $V(0)$, μ, $\tilde{\alpha}_i(0)$, $\tilde{\beta}_i(0)$, $k_i(0)$ are all parameters.

Proof. Choose the Lyapunov function candidate as follows,

$$V(t) = V_1(t) + V_2(t) + V_3(t) + V_4(t) \tag{15}$$

where

$$V_1(t) = \frac{1}{2}\bar{e}^T(t)\bar{e}(t)$$

$$V_2(t) = \frac{1}{2N}\sum_{i=1}^{N}\frac{1}{l_i}\tilde{\alpha}_i^T(t)\tilde{\alpha}_i(t)$$

$$V_3(t) = \frac{1}{2M}\sum_{i=1}^{N}\frac{1}{m_i}\tilde{\beta}_i^T(t)\tilde{\beta}_i(t)$$

$$V_4(t) = \frac{1}{2}\sum_{i=1}^{N}\frac{1}{r_i}k_i^2$$

Taking the derivative along the trajectories of the error system, we get

$$\dot{V}_1(t) = \bar{e}^T(t)\dot{\bar{e}}(t)$$

And substituting the first and second part of controller (13) to the above formula (7), one gets,

$$\dot{\bar{e}}(t) = \frac{1}{N}\sum_{i=1}^{N}F_i(x_i(t))\tilde{\alpha}_i - \frac{1}{M}\sum_{i=1}^{M}G_i(y_i(t))\tilde{\beta}_i - \frac{\sigma}{\sqrt{2^{1+\eta}}}\mathrm{sign}(\bar{e}(t))|\bar{e}(t)|^{\eta} - \frac{1}{M}\sum_{i=1}^{M}k_i(t)\bar{e}(t)$$

$$+ \frac{\sigma}{\sqrt{(2N)^{1+\eta}}}\sum_{i=1}^{N}\frac{(\|\hat{\alpha}_i\| + B_\alpha)^2}{l_i}\frac{\bar{e}(t)}{\|\bar{e}(t)\|^2} - \frac{\sigma}{\sqrt{(2M)^{1+\eta}}}\sum_{i=1}^{N}\frac{(\|\hat{\beta}_i\| + B_\beta)^2}{m_i}\frac{\bar{e}(t)}{\|\bar{e}(t)\|^2}$$

So

$$\dot{V}_1(t) = \bar{e}^T(t)\dot{\bar{e}}(t)$$

$$= \bar{e}^T(t) \cdot \frac{1}{M}\sum_{i=1}^{M}(\frac{1}{N}\sum_{i=1}^{N}F_i(x_i(t))\tilde{\alpha}_i - \frac{1}{M}\sum_{i=1}^{M}G_i(y_i(t))\tilde{\beta}_i - \frac{\sigma}{\sqrt{2^{1+\eta}}}\mathrm{sign}(\bar{e}(t))|\bar{e}(t)|^{\eta} - k_i(t)\bar{e}(t))$$

$$+ \bar{e}^T(t) \cdot \frac{1}{M}\sum_{i=1}^{M}\left(\frac{\sigma}{\sqrt{(2N)^{1+\eta}}}\sum_{i=1}^{N}\frac{(\|\hat{\alpha}_i\| + B_\alpha)^2}{l_i} - \frac{\sigma}{\sqrt{(2M)^{1+\eta}}}\sum_{i=1}^{N}\frac{(\|\hat{\beta}_i\| + B_\beta)^2}{m_i}\right)\frac{\bar{e}(t)}{\|\bar{e}(t)\|^2}$$

$$= \frac{1}{N}\bar{e}^T(t)\sum_{i=1}^{N}F_i(x_i(t))\tilde{\alpha}_i - \frac{1}{M}\bar{e}^T(t)\sum_{i=1}^{M}G_i(y_i(t))\tilde{\beta}_i - \frac{\sigma}{\sqrt{2^{1+\eta}}}|\bar{e}(t)|^{1+\eta} - \frac{1}{M}\bar{e}^T(t)\sum_{i=1}^{N}k_i(t)\bar{e}(t)$$

$$+ \frac{\sigma}{\sqrt{(2N)^{1+\eta}}}\sum_{i=1}^{N}\frac{(\|\hat{\alpha}_i\| + B_\alpha)^2}{l_i} - \frac{\sigma}{\sqrt{(2M)^{1+\eta}}}\sum_{i=1}^{N}\frac{(\|\hat{\beta}_i\| + B_\beta)^2}{m_i}$$

$$\tag{16}$$

Combine with the adaptive update laws of unknown parameters (14), we have,

$$\dot{V}_2(t) = \frac{1}{N} \sum_{i=1}^{N} \frac{1}{l_i} \tilde{\alpha}_i^T(t) \dot{\tilde{\alpha}}_i(t)$$

$$= -\frac{1}{N} \bar{e}^T(t) \sum_{i=1}^{N} F_i(x_i(t)) \tilde{\alpha}_i(t) \tag{17}$$

Similarly,

$$\dot{V}_3(t) = \frac{1}{M} \sum_{i=1}^{N} \frac{1}{m_i} \tilde{\beta}_i^T(t) \dot{\tilde{\beta}}_i(t) = \frac{1}{M} \bar{e}^T(t) \sum_{i=1}^{N} G_i(y_i(t)) \tilde{\beta}_i(t) \tag{18}$$

With the help of the feedback gain $\dot{k}_i(t)$, i.e. the third item of formula (13), one gets,

$$\dot{V}_4(t) = \sum_{i=1}^{M} \frac{1}{r_i} k_i(t) \dot{k}_i(t)$$

$$= -\sum_{i=1}^{M} \frac{\sigma}{\sqrt{(2r_i)^{1+\eta}}} |k_i|^{1+\eta} + \frac{1}{M} \sum_{i=1}^{M} (k_i + \mu_i) \bar{e}^T(t) \bar{e}(t) \tag{19}$$

Combine (16) with (17)–(19), we have

$$(\|\hat{\alpha}_i(t)\| + B_\alpha)^2 \geq (\|\hat{\alpha}_i(t)\| + B_\alpha)^{1+\eta} \geq \|\hat{\alpha}_i(t) - \alpha_i(t)\|^{1+\eta} = \left(\|\tilde{\alpha}_i\|^2\right)^{\frac{1+\eta}{2}}$$

We can also obtain,

$$\dot{V}(t) = \sum_{i=1}^{4} \dot{V}_i(t)$$

$$= \frac{1}{N} \bar{e}^T(t) \sum_{i=1}^{N} F_i(x_i(t)) \tilde{\alpha}_i - \frac{1}{M} \bar{e}^T(t) \sum_{i=1}^{M} G_i(y_i(t)) \tilde{\beta}_i - \frac{\sigma}{\sqrt{2^{1+\eta}}} |\bar{e}(t)|^{1+\eta}$$

$$+ \frac{\sigma}{\sqrt{(2N)^{1+\eta}}} \sum_{i=1}^{N} \frac{\left(\|\hat{\alpha}_i\| + B_\alpha\right)^2}{l_i} - \frac{\sigma}{\sqrt{(2M)^{1+\eta}}} \sum_{i=1}^{N} \frac{\left(\|\hat{\beta}_i\| + B_\beta\right)^2}{m_i}$$

$$- \frac{1}{N} \bar{e}^T(t) \sum_{i=1}^{N} F_i(x_i(t)) \tilde{\alpha}_i(t) + \frac{1}{M} \bar{e}^T(t) \sum_{i=1}^{N} G_i(y_i(t)) \tilde{\beta}_i(t) \tag{20}$$

$$- \sum_{i=1}^{M} \frac{\sigma}{\sqrt{(2r_i)^{1+\eta}}} |k_i|^{1+\eta} + \frac{1}{M} \sum_{i=1}^{M} (k_i + \mu_i) \bar{e}^T(t) \bar{e}(t)$$

$$= \frac{\sigma}{\sqrt{(2N)^{1+\eta}}} \sum_{i=1}^{N} \frac{\left(\|\hat{\alpha}_i\| + B_\alpha\right)^2}{l_i} - \frac{\sigma}{\sqrt{(2M)^{1+\eta}}} \sum_{i=1}^{N} \frac{\left(\|\hat{\beta}_i\| + B_\beta\right)^2}{m_i}$$

$$- \sum_{i=1}^{M} \frac{\sigma}{\sqrt{(2r_i)^{1+\eta}}} |k_i|^{1+\eta} + \frac{1}{M} \sum_{i=1}^{M} (k_i + \mu_i) \bar{e}^T(t) \bar{e}(t) - \frac{\sigma}{\sqrt{2^{1+\eta}}} |\bar{e}(t)|^{1+\eta}$$

In addition, basing on Lemma 2, one gets

$$(\|\hat{\alpha}_i(t)\| + B_\alpha)^2 \geq (\|\hat{\alpha}_i(t)\| + B_\alpha)^{1+\eta} \geq \|\hat{\alpha}_i(t) - \alpha_i(t)\|^{1+\eta} = \left(\|\tilde{\alpha}_i\|^2\right)^{\frac{1+\eta}{2}}$$

In addition, basing on the property of inequality $(b_1 + b_2 + \cdots b_n)^c \leq b_1^c + b_2^c + \cdots b_n^c$, where $0 < c < 1, b_i > 0, i = 1, 2, \cdots, n$.

$$\frac{\sigma}{\sqrt{(2N)^{1+\eta}}} \left(\sum_{i=1}^{N} \frac{\|\tilde{\alpha}_i\|^2}{l_i}\right)^{\frac{1+\eta}{2}} \leq \frac{\sigma}{\sqrt{(2N)^{1+\eta}}} \sum_{i=1}^{N} \left(\frac{\|\tilde{\alpha}_i\|^2}{l_i}\right)^{\frac{1+\eta}{2}} \leq \frac{\sigma}{\sqrt{(2N)^{1+\eta}}} \sum_{i=1}^{N} \left(\frac{\|\hat{\alpha}_i\| + B_\alpha}{l_i}\right)^2 \quad (21)$$

Similarly, we have

$$\frac{\sigma}{\sqrt{(2M)^{1+\eta}}} \sum_{i=1}^{N} \left(\frac{\|\tilde{\beta}_i\|^2}{m_i}\right)^{\frac{1+\eta}{2}} \leq \frac{\sigma}{\sqrt{(2M)^{1+\eta}}} \sum_{i=1}^{N} \left(\frac{\|\hat{\beta}_i\| + B_\beta}{m_i}\right)^2 \quad (22)$$

and

$$\sum_{i=1}^{M} \frac{\sigma}{\sqrt{(2r_i)^{1+\eta}}} |k_i|^{1+\eta} = \frac{\sigma}{\sqrt{2^{1+\eta}}} \sum_{i=1}^{M} \left(\frac{k_i^2}{\sqrt{r_i^2}}\right)^{\frac{1+\eta}{2}} \geq \frac{\sigma}{\sqrt{2^{1+\eta}}} \left(\sum_{i=1}^{M} \frac{k_i^2}{r_i}\right)^{\frac{1+\eta}{2}} \quad (23)$$

Substitute (21)–(23) to (20), we obtain,

$$\dot{V}(t) \leq -\sigma \left(\frac{1}{2N} \sum_{i=1}^{N} \frac{\tilde{\alpha}_i^T(t)\tilde{\alpha}_i(t)}{l_i}\right)^{\frac{1+\eta}{2}} - \sigma \left(\frac{1}{2M} \sum_{i=1}^{N} \frac{\tilde{\beta}_i^T(t)\tilde{\beta}_i(t)}{m_i}\right)^{\frac{1+\eta}{2}}$$

$$- \sigma \sum_{i=1}^{M} \left(\frac{1}{2} \bar{e}^T(t)\bar{e}(t)\right)^{\frac{1+\eta}{2}} - \sigma \left(\frac{1}{2} \sum_{i=1}^{M} \frac{k_i^2}{r_i}\right)^{\frac{1+\eta}{2}} + \frac{1}{M} \sum_{i=1}^{M} \mu_i \bar{e}^T(t)\bar{e}(t)$$

$$\leq -\sigma \left(\frac{1}{2} \bar{e}^T(t)\bar{e}(t) + \frac{1}{2N} \sum_{i=1}^{N} \frac{\tilde{\alpha}_i^T(t)\tilde{\alpha}_i(t)}{l_i} + \frac{1}{2M} \sum_{i=1}^{N} \frac{\tilde{\beta}_i^T(t)\tilde{\beta}_i(t)}{m_i} + \frac{1}{2} \sum_{i=1}^{M} \frac{k_i^2}{r_i}\right)^{\frac{1+\eta}{2}}$$

$$+ \frac{1}{M} \sum_{i=1}^{M} \mu_i \bar{e}^T(t)\bar{e}(t)$$

$$\leq -\sigma V^{\frac{1+\eta}{2}}(t) + \mu V(t)$$

where $\mu = \frac{2}{M} \sum_{i=1}^{M} \mu_i$.

Therefore, by Lemma 1, we can see that $V(t)$ will converge to zero in a finite time

$$t^* \leq \frac{\ln\left(1 - \frac{\mu}{\sigma} V^{1-\frac{1+\eta}{2}}(0)\right)}{\mu(\frac{1+\eta}{2} - 1)} = \frac{2\ln\left(1 - \frac{\mu}{\sigma} V^{\frac{1-\eta}{2}}(0)\right)}{\mu(\eta - 1)} \quad (24)$$

Hence, the error vector $\bar{e}(t)$ will converge to zero within t^*. Consequently, base on Lemma 1, the drive-response networks (1) and (2) are synchronized in the finite time t^*. Furthermore, the uncertain system parameter vectors $\hat{\alpha}_i(t)$, $\hat{\beta}_i(t)$ are adapted themselves to the truth values in the finite time t^*. The proof is completed.

4   Application of Finite-Time Outer Average Synchronization in Secure Communication

Based on finite-time outer average synchronization between two coupled complex dynamical networks by adaptive control, the synchronization scheme is applied to image security communication in this section. Different from the existing chaotic based image security communication schemes, image signal encryption depends on the initial value sensitivity and large key space of chaotic systems, the finite-time outer average synchronization error between two coupled complex dynamical networks were added to the process of image encryption and decryption, the newly generated encrypted signal has a large amount of data, and has the characteristics of two-dimensional spatial distribution and uneven energy distribution. When the message is transmitted, once the encrypted signal is completely intercepted, it is not easy for decrypters to use various attack means to obtain the original image signal [17], Without the key, the intercepted useful information will be 'fragmented' and difficult to decrypt. The encryption and decryption algorithm proposed in this paper requires the driving-response system to realize finite-time outer average synchronization firstly, so the following image encryption algorithm are generated after the system realizes synchronization.

Without loss of generality, Let G_N be an image of size $N \times N$, $G(x, y)(x \in [0, N-1]$, $y \in [0, N-1])$ be the gray value of image G_N at point (x, y), $P = \{1, 2, \cdots, N^2\}$ is the set of sequence points of the image, and $\overline{G}(x, y)$ is the gray value corresponding to (x, y) after encryption. The image encryption steps are as follows,

Setp 1: Reading the original image G_N, and obtaining the digital image into the matrix;

Step 2: Dividing the matrix into submatrices with size of $n \times n$, and the submatrices with RGB information are transformed by transposing horizontal flip, vertical flip, etc., to image $\tilde{G}_N$;

Step 3: Calculating the error time series $\overline{e}^i(t) = \left(\overline{e}_1^i(t), \overline{e}_2^i(t), \overline{e}_3^i(t)\right)^T$, $i = 1, 2, \cdots, n$ of the driver-response complex network composed of chaotic systems under the controller by Runge-Kutta method. However, $\overline{e}(t)$ can not be used directly to encrypt the image signal, as synchronization of the system needs time, the implementation of encryption and decryption needs to be carried out until synchronization of the system, only the average error sequence generated after t^* second is applied to image encryption. Let's denote $\overline{e}_s^i(t) = \left(\overline{e}_{s1}^i(t), \overline{e}_{s2}^i(t), \overline{e}_{s3}^i(t)\right)^T$, $i = 1, 2, \cdots, N$ be the available masking signal.

Step 4: For randomness, the following transformation is proposed,

$$\overline{e}_s^i(t) = \text{round}(\text{abs}(\overline{e}_s^i(t)) - \text{floor}(\text{abs}(\overline{e}_s^i(t)) \times 10^{15})\text{mod}256), \ i = 1, 2, \cdots, N.$$

where abs(X) means absolute value of X, floor(X) indicate taking the largest integer value less than or equal to X, and round(X) means rounding to the nearest integer value of X.

Step 5: Perform binary XOR operation on the n-th pixel grayscale value in the original image G_N and $\overline{e}_s^i(t)$ generated in step 4 to obtain the encrypted pixel value;

Step 6: Repeat steps 4 and 5 until all pixels are encrypted, resulting in an encrypted image $\tilde{G}_N$.

The decryption process is the inverse of the encryption process mentioned above, and the original image is recovered.

5 Numerical Simulations

5.1 Numerical Simulations on Finite-Time Outer Average Synchronization Between Two Coupled Networks

In the section, a numerical simulation example will be presented to illustrate the feasibility of the proposed theorem. Without of loss of generality, we choose $N = 3, M = 5$, the 3 nodes of drive network (1) are Rosslor chaotic systems, the dynamical function is as following,

$$\begin{pmatrix} \dot{x}_{i1} \\ \dot{x}_{i2} \\ \dot{x}_{i3} \end{pmatrix} = \begin{pmatrix} -x_{i2} - x_{i3} \\ x_{i1} \\ x_{i3}x_{i1} \end{pmatrix} + \begin{pmatrix} 0 & 0 & 0 \\ 0 & x_{i2} & 0 \\ 0 & 1 & -x_{i3} \end{pmatrix} \begin{pmatrix} 0 \\ p \\ q \end{pmatrix}, i = 1, 2, 3 \qquad (25)$$

When $p = 0.2$, $q = 5.7$, Rosslor system is chaotic.

And the isolate node of response network is Lorenz chaotic system,

$$\begin{pmatrix} \dot{y}_{i1} \\ \dot{y}_{i2} \\ \dot{y}_{i3} \end{pmatrix} = \begin{pmatrix} 0 \\ -y_{i1}y_{i3} - y_{i2} \\ y_{i1}y_{i2} \end{pmatrix} + \begin{pmatrix} y_{i2} - y_{i1} & 0 & 0 \\ 0 & y_{i1} & 0 \\ 0 & 0 & -y_{i3} \end{pmatrix} \begin{pmatrix} \delta \\ \theta \\ \gamma \end{pmatrix}, i = 1, 2, \cdots, 5 \quad (26)$$

Which is chaotic when the parameters are $\delta = 10, \theta = 28, \gamma = \frac{8}{3}$.

With the finite-time adaptive feedback controller (13) and parameter update laws (14), when we choose the initial values of each states in $[-5, 5]$ randomly, the parameters of feedback gains are $\eta = 0.1, \sigma = 0.2, l_i = m_i = r_i = 1, k_i(0) = 70, i = 1, 2, \cdots, 5$, $\mu_i = 0.15$. FTOAS between the above two networks will be realized with the simulation step $h = 0.005$. The initial values of unknown parameters of each node are chosen as $\alpha_{10} = [0 \ 5.18 \ 1.33]^T$, $\alpha_{20} = [0.1 \ 8.06 \ 6.16]^T$, $\alpha_{30} = [0 \ 6.78 \ 5.61]^T$, $\beta_{10} = [3.15.1814.53]^T$, $\beta_{20} = [10.65 \ 26.76.43]^T$, $\beta_{30} = [8.1 \ 25.869.29]^T$, $\beta_{40} = [9.21 \ 25.911.2]^T$, $\beta_{50} = [6.78 \ 26.57.42]^T$.

The outer coupling configuration matrices are randomly given as following,

$$A = \begin{pmatrix} -2 & 1 & 0 \\ 1 & -3 & 1 \\ 1 & 2 & -1 \end{pmatrix}, \quad B = \begin{pmatrix} -2 & 1 & 0 & 1 & 0.5 \\ 0 & -3.5 & 0 & 1 & 0.5 \\ 1 & 1.5 & -2 & 1 & 1 \\ 1 & 1 & 0 & -4 & 0.5 \\ 0 & 0 & 2 & 1 & -2.5 \end{pmatrix}$$

The simulation of FTOAS between two networks is displayed from Fig. 1, 2 and 3. Figure 1 displays the time evolution curves of the synchronization errors $\bar{e}_i(t), i = 1, 2, 3$. Each component of the error vector tends to be zero. Moreover, we can observe that the estimated parameters p,q and δ, θ, γ of drive and response networks converge to their real values $p = 0.2$, $q = 5.7$, $\delta = 10, \theta = 28, \gamma = 8/3$ showed in Fig. 2 and 3 are adjusted to fixed values in finite time t^*.

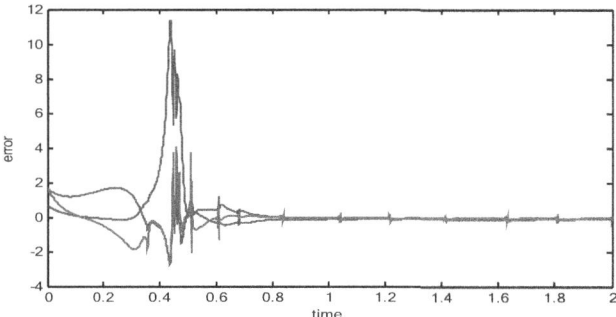

Fig. 1. Finite-time outer average error of drive-response networks when $N = 3, M = 5$

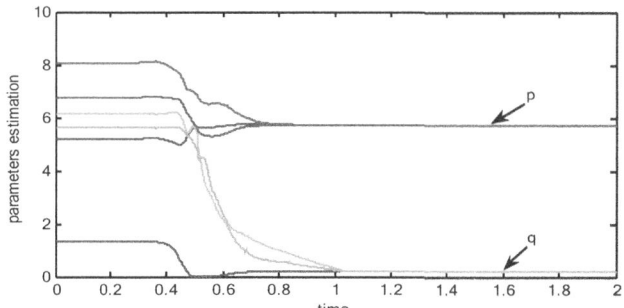

Fig. 2. Parameters estimation of drive network

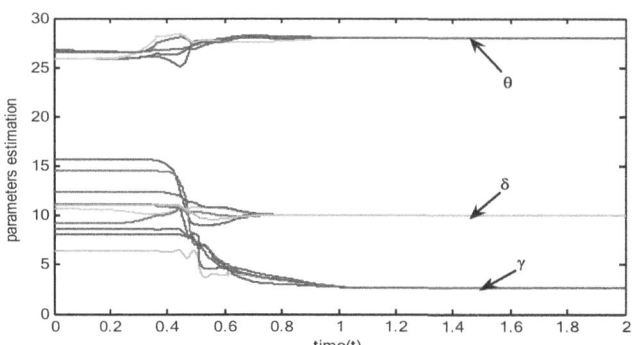

Fig. 3. Parameters estimation of response network

5.2 Numerical Simulations on Image Encryption and Decryption

The tree.png with the size of 512×512 is selected as the image to be encrypted. According to the synchronization error curve, we take the data after 10 s, that is, the complex dynamical error sequence after synchronization the system has realized finite-time outer average synchronization, to encrypt the image. The MATLAB software was used to conduct numerical simulation of the presented secure communication scheme, and the simulation results of image encryption and decryption were obtained, as shown in Fig. 4, 5, 6 and 7. The simulation results show that the image after encryption and decryption has no distortion, which shows that the image encryption and decryption scheme given is feasible.

Fig. 4. Original image

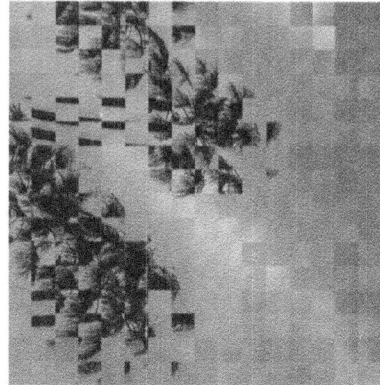

Fig. 5. Scrambled image

Fig. 6. Encryption result

Fig. 7. Decryption result

6   Conclusions

We investigated finite-time outer average synchronization problems of a class of drive-response complex dynamical networks with different dimensions of nodes and its application in image encryption and decryption in the paper. Based on finite-time stabilization theorem of nonlinear dynamical system, some useful finite-time outer average synchronization criteria has been obtained with an injected controller designed via adaptive strategy.

Also an illustrative example with MATLAB numerical simulation basing on the different dimensions of nodes was given to demonstrate the feasibility of the proposed synchronization methods. In the simulations section, the author took heterogeneity of nodes exist in many fields into account. As the whole systems' error needed for the controller designed rather than the isolate node error of each network, much larger feedback gains are needed in some cases, thanks to the use of adaptive control methods, the feedback control parameters are not too be large. We will consider the systems' fixed-time/predefined outer average synchronization with time delay and stochastic noise and its applications in the future.

Conflict of Interest. The authors declare that they have no conflict of interest.

Funding. The work is funded by the nature scientific research project of Shaanxi Education Department under Grant No. 21JK0969, the Xianyang scientific research project under Grant No. L2023-ZDYF-SF-024, and Natural Science Foundation of Xianyang Normal University under Grant No. XSYK20023.

References

1. Albert, R., Jeong, H., Barabasi, A.L.: Topology of evolving networks: local events and universality. Phys. Rev. E **85**, 5234–5237 (2000)
2. Ghaffari, A., Arebi, S.: Pinning control for synchronization of nonlinear complex dynamical network with suboptimal SDRE controllers. Nonlinear Dyn. **83**, 1003–1013 (2016)
3. Gao, P., Wang, Y., Zhao, J., et al.: Links synchronization control for the complex dynamical network. Neurocomputing **515**, 59–67 (2023)
4. Yan, L., Dai, H.: Event-triggered synchronization for second-order linear systems in complex dynamical network with time-varying topology. Int. J. Control Autom. Syst. (2024)
5. Lu, H., Hu, Y., Guo, C., et al.: Cluster synchronization for a class of complex dynamical network system with randomly occurring coupling delays via an improved event-triggered pinning control approach. J. Frankl. Inst. **357**, 2167–2184 (2020)
6. Guo, X., Li, J.: A new synchronization algorithm for delayed complex dynamical networks via adaptive control approach. Commun. Nonlinear Sci. Numer. Simul. **17**, 4395–4403 (2012)
7. Tan, M., Tian, W.: Finite-time stabilization and synchronization of complex dynamical networks with nonidentical nodes of different dimensions. Nonlinear Dyn. **79**, 731–774 (2015)
8. Zhang, L., Lei, Y., Wang, Y., et al.: Matrix projective synchronization for time-varying disturbed networks with uncertain nonlinear structures and different dimensional nodes. Neurocomputing **311**, 11–23 (2018)
9. Hao, X., Li, J.: Stochastic synchronization for complex dynamical networks with time-varying couplings. Nonlinear Dyn. **80**, 1357–1363 (2015)
10. Yang, X., Cao, J.: Finite-time stochastic synchronization of complex networks. Appl. Math. Model. **34**, 3631–3641 (2010)
11. Sun, Y., Li, W., Zhao, D.: Finite-time stochastic outer synchronization between two complex dynamical networks with different typologies. Chaos **22**, 023152 (2012)
12. Zhao, H., Li, L., Peng, H.: Mean square modified function projective synchronization of uncertain complex network with multi-links and stochastic perturbations. Eur. Phys. J. B **88** (2015). https://doi.org/10.1140/epjb/e2014-50577-2
13. Yan, L., Li, J.: Outer average synchronization between two coupled networks with different numbers of nodes. IMA J. Math. Control. Inf. **36**, 1136–1149 (2019)
14. Hu, C., Jiang, H.: Pinning synchronization for directed networks with node balance via adaptive intermittent control. Nonlinear Dyn. **80**, 295–307 (2015)

15. Liu, X., Daniel, H., Song, Q., et al.: Finite/Fixed-time pinning synchronization of complex networks with stochastic disturbances. IEEE Trans. Cybern. **99**, 1–6 (2018)
16. Xie, Q., Si, G., Zhang, Y., et al.: Finite-time synchronization and identification of complex delayed networks with Markovian jumping parameters and stochastic perturbations. Chaos Solitons Fractals **86**, 35–49 (2016)
17. Zhang, H., Liu, X., Shen, X., et al.: Intermittent impulsive synchronization of hyperchaos with application to secure communication. Asian J. Control **15**, 1686–1699 (2013)

Classification of In-Situ Solar Wind Data Measured by Solar Orbiter/SWA-PAS and HIS Using Machine Learning

Liang Zhao[1(✉)] [iD], Henry Han[2(✉)] [iD], Susan T. Lepri[1(✉)], and Ryan Dewey[1(✉)]

[1] Department of Climate and Space Sciences and Engineering,
University of Michigan, Ann Arbor, MI 48109, USA
{lzh,slepri,rmdewey}@umich.edu
[2] Department of Computer Science, School of Engineering and Computer Science,
Baylor University, Waco, TX 76798, USA
henry_han@baylor.edu

Abstract. Connecting in-situ solar wind properties with their source regions on the Sun has long been one of the unsolved questions in heliophysics. This challenge can now be addressed using modern AI/ML techniques and big data analysis algorithms. In this work, we apply state-of-the-art AI/ML technology on in-situ solar wind measurements made by the Heavy Ion Sensor (HIS) and Proton and Alpha Particle Sensor (PAS) onboard the recent Solar Orbiter mission (launched in 2020) to classify different types of solar wind. These data-driven classifications may provide insights into their coronal origins and could significantly help heliophysicists in understanding the long-standing question of where the slow solar wind originates.

Keywords: Solar wind · Classification · Machine Learning · Heliophysics

1 Introduction

Solar wind is crucial to space weather science and forecasting because the properties of the solar wind plasma affect the local conditions in the entire Heliosphere, and largely determine the propagation, arrival time and geo-effectiveness of solar eruptions. In addition, solar wind plasma is inextricably tied to the thermal properties of the inner corona where it is accelerated, heated, and ionized. For example, coronal holes (CHs), where the fast wind is accelerated from, are regions where the plasma electron density and temperature are lower than in their surroundings, and consequently the wind originating from CHs is usually characterized by overall lower plasma ionization and heavy ion charge state ratios (e.g., O^{7+}/O^{6+} or C^{6+}/C^{5+}) [1]. Figure 1 shows polar plots of Ulysses'first (18 June 1993–10 January 1997) and third orbit (30 October 2005–30 June 2009) observations during solar minima: fast and less ionized wind was dominant in high latitude regions where polar CHs were prevalent.

H. Han (Ed.): SDSC 2024, CCIS 2158, pp. 183–198, 2024.
https://doi.org/10.1007/978-3-031-67871-4_13

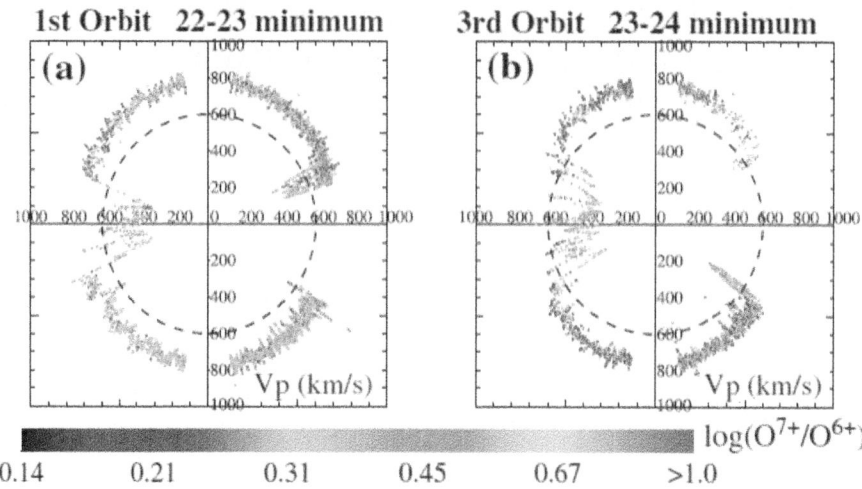

Fig. 1. Polar plots of solar wind proton speed (color coded with O^{7+}/O^{6+} ratio) from Ulysses' first (a) and third (b) orbits. Time goes in clockwise direction, starting from the bottom right quadrants. Figure adapted from [6]

On the contrary, no such direct connection can be found between the slow-speed solar wind and a specific source region in the Sun: the slow solar wind might be originating from anywhere outside polar CHs, and even from their edges. Figure 1 shows that in the low latitude regions the solar wind speed are mostly slow or intermediate and the ionization level (indicated by O^{7+}/O^{6+} ratio) is higher than in the fast wind. The sources of these slow and hot wind has been variously suggested to be multiple types of coronal regions, including the periphery of active regions (ARs, e.g., [2,3]), helmet streamers [4,5,7], CH boundaries [8,9], or pseudostreamers [10]. The latter structure consists of two sets of closed magnetic loops that separate open flux with the same polarity [11,12], and they were abundant between 2007 and 2009 [13,14]. The solar wind originating from pseudostreamer regions has been suggested to have slow to intermediate speeds and to be highly ionized (high O^{7+}/O^{6+} ratio, [6]). Besides these origins, the slow-speed wind can also come from low-latitude CHs. For example, slow solar wind ($V < 600\,\mathrm{km/s}$) observed by ACE around the ecliptic plane has been associated with low-latitude CHs because its variability, composition, and properties are very similar to CH-associated fast wind [4,7]. Now the question is, besides the clear association between the fast wind and polar CHs, the solar origins of the low to intermediate speed wind still remain a puzzle.

To solve this puzzle, solar wind heavy ion composition measurements are the key factor that we need to rely on to link the solar wind from the heliosphere to its solar sources. The important role that the heavy ion charge state and element abundance play in understanding the solar wind and the heliosphere is twofold. First, they are unique tools to connect the in-situ solar wind properties to their origins in the corona. As the solar wind is accelerated, the plasma ionization and recombination rates are proportional to the electron density, which rapidly decreases with distance. At a critical height where the electron density is so low that the two processes effectively stop [1] and the ionization state of an element remains unaltered as the solar wind propagates into the heliosphere (so-called "freeze-in" process, Fig. 2); hence solar wind ion composition provides a direct measure of the thermal properties in the corona [4,5,15,16]. In addition, element abundance ratios can be affected by fractionation processes in the upper chromosphere resulting in a significant enhancement of the abundance of elements with low First Ionization Potential (FIP < 10 eV) in closed magnetic, field structures over their values in photosphere [7,17–19]. This is the so-called FIP effect (Fig. 3, Laming 2015). Such fractionation is directly inherited by the wind plasma and once the wind plasma is accelerated away, the abundance values remain unaltered and thus maintain record of the composition of the source region (e.g., [20]).

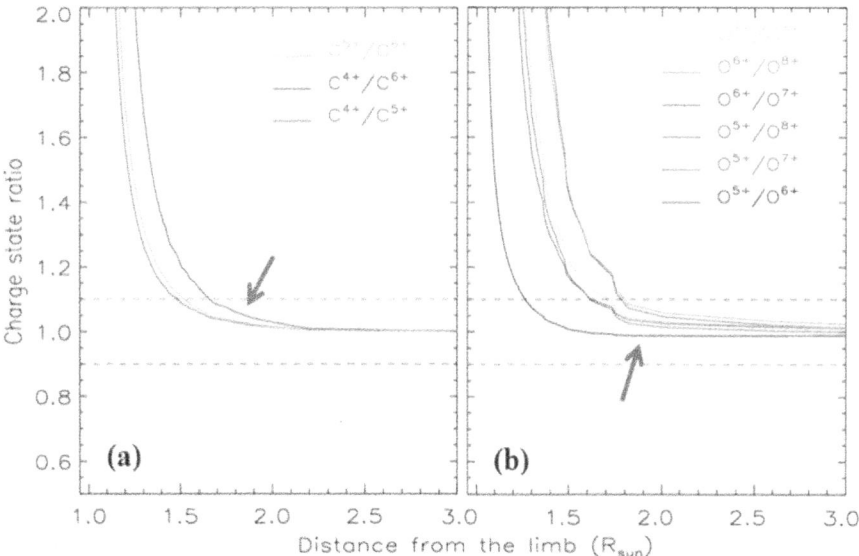

Fig. 2. Evolution of charge state ratios of Carbon (a) and Oxygen (b) as a function of distance from the solar limb (at 1 R$_s$) calculated by an Ionization model [21]. Red arrows point to where the charge states of C and O are frozen-in. Figure adapted from [21] (Color figure online)

Second, in-situ anomalous composition data are tightly linked to the process of CME eruptions and can yield important insights into the characteristics of the CME source environment. [22] presented the first systematic search for hot material inside ICMEs observed by ACE/SWICS and found that more than 50% of ICMEs exhibited long periods (>20 h) of enhanced $Fe^{>16+}/Fe_{total}$ (Fig. 4).

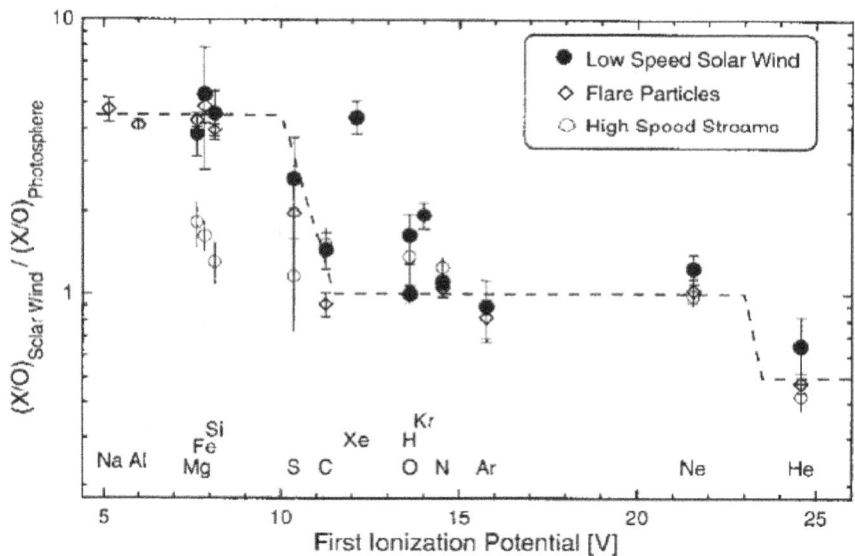

Fig. 3. Elemental abundances as a function of FIP in the slow wind and fast wind. Abundances are given relative to O and normalized to the photosphere abundances (Anders and Grevesse, 1989). Figure adapted from [17]

Later work by Gruesbeck et al. (2011) showed that elevated Fe charge states are actually present in more than 95% of ICMEs. These elevated charge states (C, O, Si, Fe) indicate enhanced electron densities and temperatures most likely linked to plasma heating due to the deposition of magnetic energy during reconnection. [23] also reported a number of events where unusually cold material (Fe^{4-7+}, O^{2-4+}, C^{2-3+}) was also present in in-situ measurements of ICMEs; this indicates the presence of low electron temperatures in the source region, likely related to prominence material.

Since solar wind heavy ions play a crucial role as significant test particles and respond uniquely to the heliospheric environment surrounding them, understanding the properties of the solar wind heavy ions composition is imperative for tracing heliospheric structures to their sources on the Sun or local sources in interplanetary space. The Solar Orbiter mission [24] is equipped with a Solar Wind Analyser (SWA, [25]), comprising the Proton and Alpha Particle Sensor (PAS) and the Heavy Ion Sensor (HIS, [26]). The integration of these new instruments opens a new era of modern solar wind heavy ion in-situ observations, succeeding the pioneering missions of Ulysses and ACE. Now we have a unique opportunity to fully harness the scientific potential of solar wind composition measurements by Solar Orbiter, allowing us to gain deeper insights into our understanding of the inner heliosphere and beyond.

It is the purpose of this paper to apply Machine Learning clustering algorithms on the recent Solar Orbiter observations during 2022, in order to category

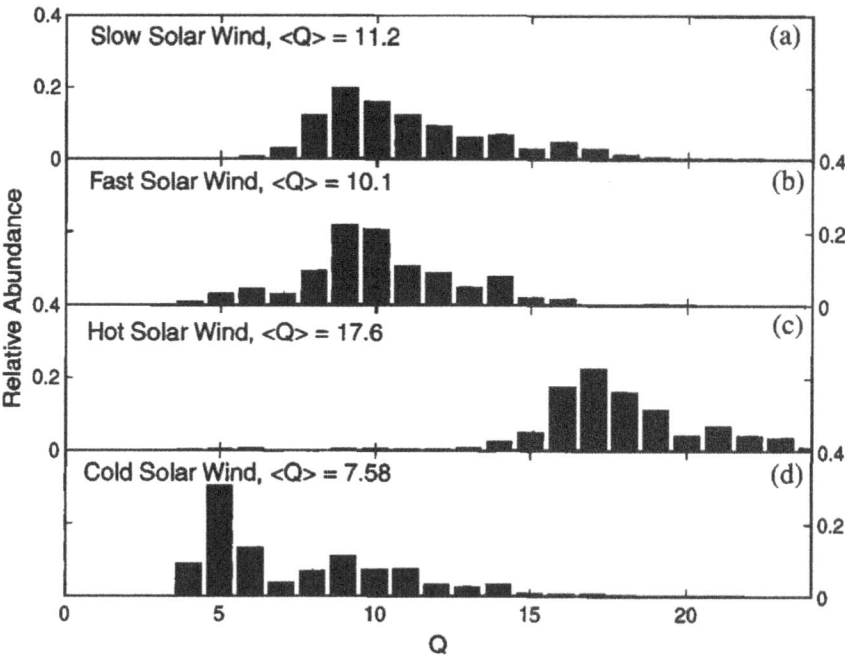

Fig. 4. Examples of Fe charge state distributions. (a) A distribution for typical slow solar wind with an average charge state corresponding to a coronal collisional equilibrium electron temperature of T = 1.4 MK (0000–0200 UT, September 7, day 250, 1998). (b) A distribution for fast solar wind with a corresponding temperature of T = 1.2 MK (0000–0200 UT, July 24, day 205, 1998). (c) A distribution for hot solar wind corresponding to coronal collisional equilibrium electron T = 7.3 MK (2000–2300 UT, September 25, day 268, 1998). (d) A cold solar wind charge distribution with T = 0.6 MK (000-0200 UT, September 25, day 268, 1998). Note that the distributions in Figs. 1b and 1c were obtained on the same day. Figure adapted from [22]

the data into multiple clusters and then to understand the differences between these clusters in terms of their in-situ physical properties, and their possible difference in the coronal origins. In Sect. 2, we introduce the Solar Orbiter data we use in this work. Section 3 describes the ML feature selection process that we apply on the solar orbiter data. In Sect. 4, we discuss the ML clustering procedure and present the clustering results. In Sect. 5, we investigate the differences in the in-situ properties of these different clusters of solar wind as identified by the ML algorithm, and discuss the implications for their coronal origins. Finally, in Sect. 6, we summarize our results, discuss the significance of this work, and outline future plans.

2   Solar Orbiter Observations

With the launch of Solar Orbiter in 2020 [24], new measurements of solar wind, including heavy ion properties, are now available in the inner heliosphere. These measurements, from the Solar Wind Analyzer (SWA, [25]) allow us to extender our previous analysis (e.g., [4,5] into the most recent solar cycle. In particularly, the SWA suite is comprised of the Proton and Alpha Sensor (PAS), the Electron and Alpha Sensor (EAS-double check acronym) and the Heavy Ion Sensor (HIS, [26]), and provides the opportunity to investigate the properties of the slow-speed solar wind from 2022, close to the solar maximum of cycle 25.

The in-situ solar wind measurements we used in the work as input for the ML clustering models are: 10-min resolution of O^{7+}/O^{6+}, O^{6+}/O^{4+}, C^{6+}/C^{5+}, abundance ratio Fe/O, and average charge state of Oxygen, from Solar Orbiter/HIS, during Jan. 2022 to Apr. of 2024. And proton speed and density data as measured by Solar Orbiter/PAS, averaged into the 10-min winder to match with HIS data.

3   Feature Selection to Rank the Importance of the Solar Wind Parameters of Solar Orbiter Observation

Feature selection is a crucial process in ML/AI data analysis. The goal of this procedure is to rank the input features (variables) by their importance, so that to enhance the interpretability of the data, decrease computational complexity, improve model performance, and reduce possible overfitting. In this study, we apply ML feature selection on the Solar Orbiter/PAS and HIS measurements duirng 2022. By discerning which variables are essential, we build an interpretable and efficient ML/AI systems by exploiting the most influential variables. Figure 5 we utilize extremely randomized trees (extra-tree), random forests, and support vector machines (SVM) to rank the feature importance for the seven Solar Orbiter SWA measurements in 2022, including proton speed and density made by PAS, and O^{7+}/O^{6+}, C^{6+}/C^{5+}, C^{6+}/C^{4+}, average charge state of $O(<Q>_O)$, and Fe/O density ratio made by HIS. Across the three models we try, the O^{7+}/O^{6+} emerges as the most critical variable, followed by the $<Q>_O$ and proton speed. The Fe/O ratio turns out to be the least important feature compared to the other six, which is actually consistent with the study by Stakhiv et al. (2015, 2016) that shows the insensitivity of the Fe/O across intermediate and fast speed solar wind.

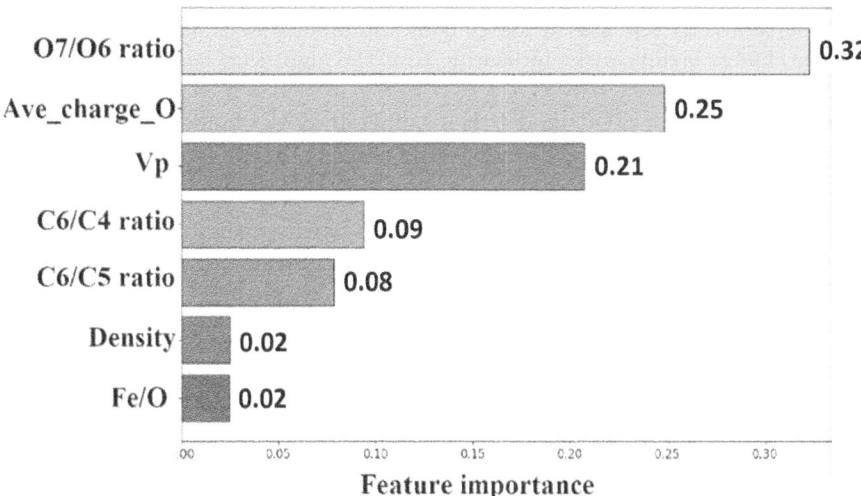

Fig. 5. Optimization of the solar wind in-situ SolO measurements by three ML algorithms (extra-tree, random forests, and SVM)

4   ML Clustering and Results

We proposed a self-supervised learning model, Solar-CDC, to handle high-performance unsupervised learning for solar wind data. It integrates the following basic ideas with a transformer for input data encoding.

Generate Pseudo Labels: Initially, the algorithm generates pseudo labels for the training data. These labels are based on the inherent structure and features of the data itself, rather than external annotations.

Use Pseudo Labels for Training: These pseudo labels are then used as if they were real labels to train a model. This step involves treating the pseudo-labeled data in a manner similar to a supervised learning problem, where the model learns to predict these labels.

Reassign Pseudo Labels: As the model learns and improves, the pseudo labels are periodically reassigned. This reassignment is based on the model's evolving understanding of the data, ensuring that the labels remain relevant and accurate.

For pseudo-label generation, we used the K-Means algorithm. K-Means is a widely used partitioning method in data analysis for dividing a dataset into k distinct clusters based on similarity. K-Means is to minimize the following Within-Cluster Sum of Squares (WCSS)

$$\text{WCSS} = \sum_{j=1}^{k} \sum_{x_i \in C_j} d(x_i, \mu_j)^2, \tag{1}$$

where C_i is the set of points in cluster i, and μ_i is the mean of points in C_i. The algorithm iteratively repeats the assignment and update steps until the centroids no longer change significantly, indicating that the algorithm has converged to a solution.

The proposed solar-CDC model then builds upon the fundamental learning principles with an emphasis on conducting clustering in the latent space generated by a transformer.

Encoder for Latent Representation: The process begins with an encoder, implemented as a transformer model in this study. It transforms the high-dimensional data into a lower-dimensional latent representation, capturing the essential characteristics of the data.

Cluster Latent Representation: The latent representation is then clustered using an unsupervised algorithm, grouping the data based on similarities identified in the latent space.

Creation of New Dataset with Pseudo Labels: The cluster assignments from the latent representation are used as new pseudo labels, effectively creating a new dataset where each data point is labeled based on its assigned cluster.

Addition of a Classification Head: A classification head is added to the encoder. This head is trained to predict the pseudo labels as if they were true labels, adhering to a supervised learning paradigm. This part of the process is known as the 'regular step'.

Transformer. We briefly introduce the transformer as follows for the sake of understanding. The transformer encoder leverages self-attention to capture relationships in an input sequence, which is solar wind data in our study, with the following steps.

1. **Input Representation:** Let $\mathbf{X} \in \mathbb{R}^{n \times d}$ be the input matrix, where n is the sequence length and d is the dimensionality of each input vector.
2. **Multi-Head Self-Attention:** Attention scores are computed as:

$$\text{Attention}(\mathbf{Q}, \mathbf{K}, \mathbf{V}) = \text{softmax}\left(\frac{\mathbf{Q}\mathbf{K}^{\top}}{\sqrt{d_k}}\right)\mathbf{V} \tag{2}$$

with $\mathbf{Q} = \mathbf{X}\mathbf{W}_Q$, $\mathbf{K} = \mathbf{X}\mathbf{W}_K$, $\mathbf{V} = \mathbf{X}\mathbf{W}_V$, and multiple heads combined as:

$$\text{MultiHead}(\mathbf{Q}, \mathbf{K}, \mathbf{V}) = \text{Concat}(\text{head}_1, \ldots, \text{head}_h)\mathbf{W}_O \tag{3}$$

where each head_i uses different learned projections.
3. **Feed-Forward Network:** Each position uses:

$$\text{FFN}(x) = \max(0, x\mathbf{W}_1 + b_1)\mathbf{W}_2 + b_2 \tag{4}$$

4. **Layer Normalization and Residual Connections:** Normalization and residual connections are applied as:

$$\mathbf{Z}_1 = \text{LayerNorm}(\mathbf{X} + \text{MultiHead}(\mathbf{Q}, \mathbf{K}, \mathbf{V})) \tag{5}$$

$$\mathbf{Z}_2 = \text{LayerNorm}(\mathbf{Z}_1 + \text{FFN}(\mathbf{Z}_1)) \tag{6}$$

Stacking these layers allows the model to learn complex dependencies, useful for tasks like unsupervised learning for solar wind data to seek meaningful clustering in the proposed solar-CDC model (Fig. 6).

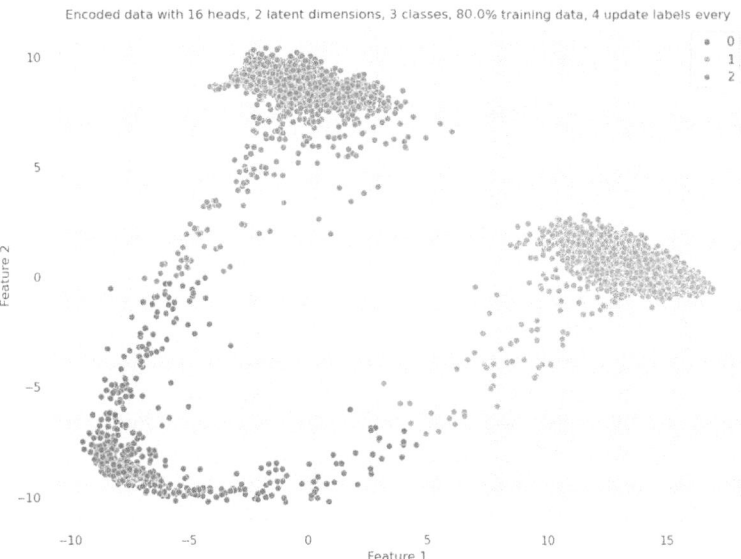

Fig. 6. The three clusters detected in the latent space by the encoder (transformer) from Solar-CDC.

5   Physical Properties of the Clusters

After the solar wind observations are clustered into different categories by the ML models, the physical properties (such as proton dynamic properties and heavy ion compositions) of each cluster can be investigated and compared. Figure 7 summarize a comparison of these physical properties in the four clusters of solar wind as identified by the proposed Solar-CDC model (Fig. 6). The observations shown in Fig. 7 are 10-min measurements made by Solar Orbiter/HIS and 10-min averages of the 1.6-min data made by PAS during 2022.

Figure 7a displays the comparison of the O^{7+}/O^{6+} and the proton speed; these two observables are in general anti-correlated when large ICMEs intervals are excluded. This anti-correlation has been empirically confirmed by decades of solar wind in-situ measurements, such as Ulysses [27], ACE [4], and Solar Orbiter [28]; and also theoretically interpreted by the magnetic interchange reconnection for the solar wind plasma acceleration process in the corona (e.g. [28]). On the other hand, data analysis studies for the in-situ properties of the ICME plasma show that during long ICME intervals, the heavy ion charge state, such as O^{7+}

[53] and $Fe^{>16+}$ [54] tend to be enhanced. Based on these statistical study, an empirical threshold using the O^{7+}/O^{6+} and proton speed to identify the ICME intervals is find (gold line in Fig. 7a Richardson and Cane 2004). Data points that are above this threshold in 7a are likely to be ICME plasmas. The four clusters of the solar wind mostly still follow this anti-correlation, with some of the blue points (Cluster 2) are distributed above the gold line, which implies that some of the solar wind in Cluster 2 may be strongly suspected to be associated with ICMEs.

Figure 7b shows the histograms of O^{7+}/O^{6+} ratio in these four clusters. The most interesting feature of this plot is that Cluster 2 (blue) and Cluster 3 (black) are distinctly separated on either side of the vertical dotted line where O^{7+}/O^{6+} = 0.145. This separating point of O^{7+}/O^{6+} = 0.145 has been suggested by [4,5] to distinguish the solar wind associated with coronal hole regions from those come from the outside of coronal holes, based on observations of ACE mission in the solar cycle 23, using traditional data analysis method. [39] employed a different set of data analysis focusing on analyzing the fast and slow wind interfaces, coincidentally, they also find that this find a separating point between the fast and slow streams at this value. Note that these previous work do not use ML/AI technologies. In this work, we show that a clustering result provided by solar-CDC ML model strongly suggests that some very distinguishable groups of solar wind could be separated at O^{7+}/O^{6+} = 0.145. The consistence of the results by the non-ML and ML analysis confirms: 1) the O^{7+}/O^{6+} = 0.145 threshold is indeed a critical point, with the solar wind data on either side of this value exhibiting distinct physical features; 2) our ML model indeed uncovers some very essential features from the data that has physical magnificence.

Besides the very distinguished Cluster 2 and 3, Fig. 7b also shows that a large cluster (Cluster 1) that possesses about double of the data points as in Cluster 2, has a O^{7+}/O^{6+} distribution in between of the Cluster 2 and 3, with a slight skew towards the lower value side, indicating that this cluster might be a combination of the both coronal-hole associated wind (as Cluster 2) and non-coronal-hole associated wind (as Cluster 3), and the current solar-CDC model did not categorize this Cluster 1 well. Also the Cluster 4 (red) which has the minimum data points, has a wide spread of O^{7+}/O^{6+} value range, indicating that the solar-CDC model did not classify this type well.

Figure 7c represents the histograms of proton speed in the four clusters, and confirms with our understanding as learned from Fig. 7b: Cluster 2 are non-coronal-hole associated slow speed solar wind (V < 450 Km/s according to the blue histogram) and Cluster 3 is a coronal-hole-associated wind. However, we also notice that the proton speed distribution in Cluster 3 has two peaks, one is around the fast speed range (V $\sim$ 500 Km/s), and the other smaller one is around 380 Km/s. This bi-model speed feature of the coronal-hole-associated wind is exactly what we have found in previous study with ACE observations [4,7]. The fast subgroup of the Cluster 3 might be associated to large coronal holes; and the slower subgroup might be associated with equatorial, small-sized and isolated coronal holes.

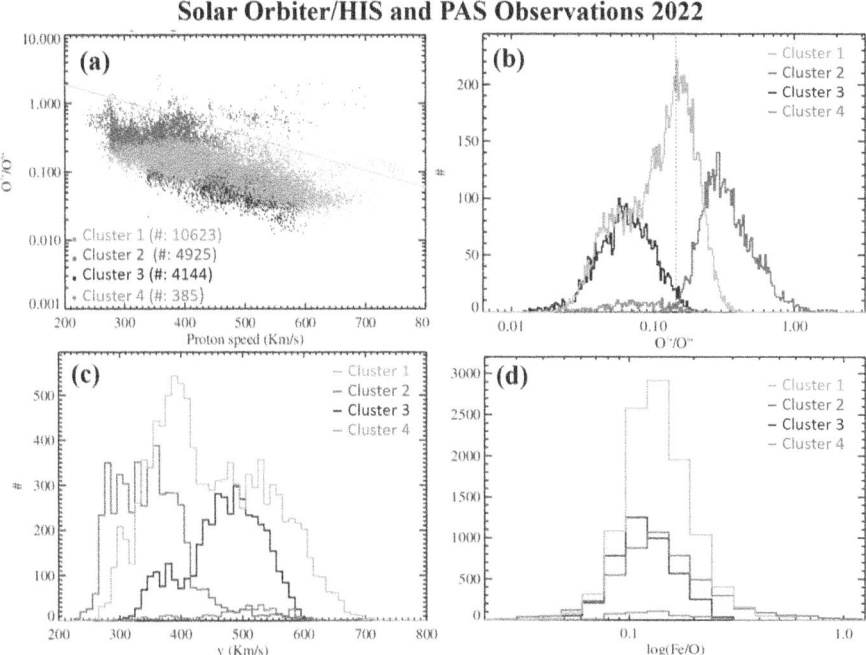

Fig. 7. Solar Orbiter HIS and PAS observations during the year of 2022. (a) Scattered plot of O^{7+}/O^{6+} versus proton speed ratios. Four clusters are color coded in green, blue, black and red with the numbers of the 10-min data point labeled in the legend. Gold line is the ICME empirical threshold given by Richardson & Cane 2004. (b) Histograms of O^{7+}/O^{6+}, (c) proton speed, and (d) Iron to Oxygen abundance ratio in the four clusters as identified by the Solar-CDC model. Vertical dotted line marks where $O^{7+}/O^{6+} = 0.145$. (Color figure online)

Figure 7d shows that the histograms of Fe/O abundance ratio in these four clusters are overlapped. This result is actually not very surprising. As reported by [8], Fe/O ratio could be vary similar in the coronal-hole and coronal-hole boundary associated wind, the only wind that has slightly enhanced Fe/O is the very slow, streamer associated wind in the equatorial region. The overlapping of the Fe/O histograms in these four clusters implies that our current ML model do not capture the streamer-associated high Fe/O ratio wind well. However, on the other hand, according to the feature selection result in Fig. 5, Fe/O ratio is ranked the lowest importance, indicating that this parameter is not very efficient to be used to identify the features in the solar wind data. That could also be a reason why the outcome properties of Fe/O ratio in the four clusters are not distinguished well.

Figure 8 shows the physical properties of another clustering result using our solar-CDC deep clustering model on the extended Solar Orbiter dataset from Jan. 2022 to April 2023. The outcome is 3 clusters: Cluster 1 (red) possesses

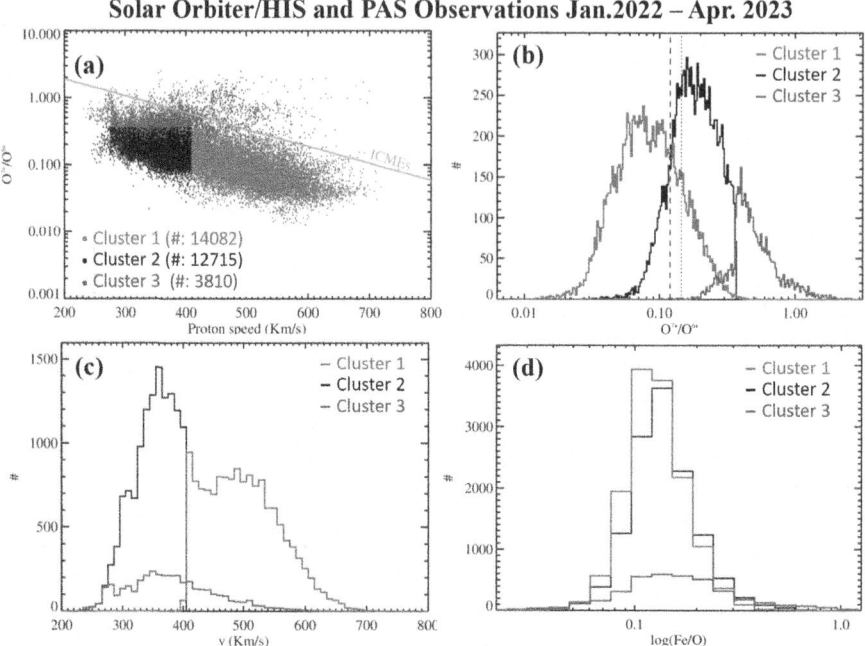

Fig. 8. Solar Orbiter HIS and PAS observations during Janunary 2022 to April 2023. (a) Scattered plot of O^{7+}/O^{6+} versus proton speed ratios. Three clusters are color coded in red, black, and blue with the numbers of the 10-min data point labeled in the legend. Green line is the ICME empirical threshold given by Richardson & Cane 2004. (b) Histograms of O^{7+}/O^{6+}, (c) proton speed, and (d) Iron to Oxygen abundance ratio in the clusters as identified by the Solar-CDC model. Vertical lines marks where $O^{7+}/O^{6+} = 0.145$ (dotted), and 0.12 (dashed). (Color figure online)

relatively low O^{7+}/O^{6+} ratio (Fig. 8b), high proton speed (> 400 Km/s, 8c); Cluster 2 (black) has relatively high O^{7+}/O^{6+} ratio (Fig. 8b) and slow proton speed (>400 Km/s, 8c); and Cluster 3 (blue) is the hotest group with highest O^{7+}/O^{6+} ratio (Fig. 8a and b), however wide spread speed range. These features implies that Cluster 1 might be the coronal-hole associated fast wind; Cluster 2 is likely the non-coronal-hole, streamer-associated slow wind; and some of the Cluster 3 plasma could be ICMEs. The overlapping of the Fe/O histogram (Fig. 8d) is again could because the nature of Fe/O itself that is not very effective to distinguish the types of solar wind, or could be attributed to the performance of our ML model.

The separation of Cluster 2 and 3 is at $O^{7+}/O^{6+} = 0.12$ (dashed line in 8b), very close to the threshold we discussed in Fig. 7b, ensuring the robustness of this separation point between different solar wind types.

6   Discussion and Conclusion

This coincidence of finding $O^{7+}/O^{6+} = 0.145$ as the separation point for coronal-hole and non-coronal-hole associated wind by our ML model indicates at least two benefits of using ML/AI to classify the solar wind types. 1) Due to their data-driven nature, ML/AI algorithms is a more objective and efficient approach to solar wind classification, while traditional methods often involve subjective or arbitrary decisions. Therefore, the result of ML/AI is closer to the nature of the data, with minimal human intervention. 2) Because of 1), ML/AI is excelled at extracting scientific insights from data more efficiently than traditional model-driven methods. Being adaptive and learning directly from data patterns make ML/AI more easily to unveil scientific characteristic hidden in the various solar wind datasets, especially extensively large ones, like Solar Orbiter.

The proposed Solar-CDC model demonstrates good latent meaningful clustering discovery capabilities. However, it sometimes exhibits overfitting by finding clusters limited to some important features such as velocity. It is possible that the transformer in the Solar-CDC does not capture the dynamics of solar wind time series data.

To improve the Solar-CDC model and address overfitting, we plan to enhance data representation by incorporating additional features, apply regularization techniques, adjust the model architecture by experimenting with hybrid models, fine-tune the attention mechanism, augment the training data, implement time series-specific techniques, optimize hyperparameters, use cross-validation, and post-process clusters based on domain knowledge. These steps will help the model capture the complex dynamics of solar wind data and reduce overfitting, leading to more accurate and generalizable clustering results.

Acknowledgements. L.Z. is supported by NASA Grants 80NSSC21K0579, 80NSSC22K1015, NSF SHINE grant 2229138, and NSF Early Career grant 2237435. H.H is supported by the McCollum endowed chair startup fund of Baylor University.

References

1. Owocki, S.P., Holzer, T.E., Hundhausen, A.J.: The solar wind ionization state as a coronal temperature diagnostic. Astrophys. J. **275**, 354–366 (1983)
2. Ko, Y.-K., Raymond, J.C., Zurbuchen, T.H., Riley, P., Raines, J.M., Strachan, L.: Abundance variation at the vicinity of an active region and the coronal origin of the solar solar wind. ApJ **646**(2), 1275–1287 (2006)
3. Liu, S., Su, J.T.: Multi-channel observations of plasma outflows and the associated small-scale magnetic field cancellations on the edges of an active region. Astrophys. Space Sci. **351**(2), 417–425 (2014)
4. Zhao, L., Zurbuchen, T.H., Fisk, L.A.: Global distribution of the solar wind during solar cycle 23: ACE observations. Geophys. Res. Lett. **36**, L14104 (2009)
5. Zhao, L., Landi, E., Zurbuchen, T.H., Fisk, L.A., Lepri, S.T.: The evolution of 1 AU equatorial solar wind and its association with the morphology of the heliospheric current sheet from solar cycles 23 to 24. Astrophys. J. **793**, 44, 8p. (2014) https://doi.org/10.1088/0004-637X/793/1/44

6. Zhao, L., Gibson, S.E., Fisk, L.A.: Association of solar wind proton flux extremes with pseudostreamers. J. Geophys. Res. Space Phys. **118**, 2834–2841 (2013)
7. Zhao, L., et al.: On the relation between the in-situ properties and the coronal sources of the solar wind. Astrophys. J. **846**(2), 135 (2017)
8. Stakhiv, M., Landi, E., Lepri, S.T., Oran, R., Zurbuchen, T.H.: On the origin of mid-latitude fast wind: challenging the two-state solar wind paradigm. Astrophys. J. **801**, 100 (2015)
9. Stakhiv, M., Lepri, S.T., Landi, E., Tracy, P., Zurbuchen, T.H.: On solar wind origin and acceleration: measurements from ACE. Astrophys. J. **829**(2), 117 (2016)
10. Owens, M.J., Crooker, N.U., Lockwood, M.: Solar cycle evolution of dipolar and pseudostreamer belts and their relation to the slow solar wind. J. Geophys. Res. Space Phys. **119** (2014). https://doi.org/10.1002/2013JA019412
11. Hundhausen, A.J.: An interplanetary view of coronal holes. In: Coronal Holes and High Speed Wind Streams, pp. 225–329 (1977)
12. Wang, Y.-M., Sheeley Jr., N.R., Rich, N.B.: Coronal pseudostreamer. ApJ **658**, 1340 (2007)
13. Abramenko, V., Yurchyshyn, V., Linker, J., Mikić, Z., Luhmann, J., Lee, C.: Low-latitude coronal holes at the minimum of the 23rd solar cycle. ApJ **712**(2), 813–818 (2010)
14. de Toma, G., Gibson, S.E., Emery, B.A., Arge, C.N.: Evolution of coronal holes and implications for high-speed solar wind during the minimum between cycles 23 and 24. In: SOHO-23 Proceeding, vol. 217 (2010)
15. Zhao, L., Landi, E., Fisk, L.A., Lepri, S.T.: The coherent relation between the solar wind proton and speed and O7+/O6+ ratio and its coronal sources. In: AIP Conference Proceedings, vol. 1720, p. 020007 (2016)
16. R. von Steiger, Zurbuchen, T.H.: Polar coronal holes during the past solar cycle: ulysses observations. J. Geophys. Res. (Space Phys.) **116**, A01105 (2011)
17. Geiss, J.: Constraints on the FIP mechanisms from solar wind abundance data. Space Sci. Rev. **85**(1–2), 241–252 (1998). https://doi.org/10.1023/A:1005198416520
18. Schwadron, N.A., Fisk, L.A., Zurbuchen, T.H.: Elemental fractionation in the slow solar wind. Astrophys. J. **521**(2), 859–867 (1999). https://doi.org/10.1086/307575
19. Von Steiger, R., et al.: Composition of Quasi-Stationary Solar Wind Flows from Ulysses/Solar Wind Ion Composition Spectrometer. J. Geophys. Res. **105**, 27217–27238 (2000). https://doi.org/10.1029/1999JA000358
20. von Steiger, R., Zurbuchen, T.H.: Solar metallicity derived from in-situ solar wind composition. Astrophys. J. **816**, 13 (2016)
21. Landi, E., Gruesbeck, J.R., Lepri, S.T., Zurbuchen, T.H., Fisk, L.A.: Charge state evolution in the solar wind. II. Plasma charge state composition in the inner corona and accelerating fast solar wind. Astrophys. J. **761**, 48 (2012)
22. Lepri, S.T., Zurbuchen, T.H., Fisk, L.A., Richardson I.G., Cane, H.V., Gloeckler, G.: Iron charge distribution as an identifier of interplanetary coronal mass ejections. J. Geophys. Res. **106**(A12), 29,231–29,238 (2001)
23. Lepri, S.T., Zurbuchen, T.H.: Direct observational evidence of filament material within interplanetary coronal mass ejections. ApJL **723**, L22 (2010)
24. Müller, D., St. Cyr, O.C., Zouganelis, I., et al.: The solar orbiter mission - science overview. A&A **642**, A1 (2020). https://doi.org/10.1051/0004-6361/202038467
25. Owen, C.J., Bruno, R., Livi, S., et al.: The solar orbiter solar wind analyser (SWA) suite. A&A **642**, A16 (2020). https://doi.org/10.1051/0004-6361/201937259
26. Livi, S., Lepri, S.T., Raines, J.M., et al.: First results from the solar orbiter heavy ion sensor. A&A **676**, A36 (2023). https://doi.org/10.1051/0004-6361/202346304

27. Gloeckler, G., Zurbuchen, T.H., Geiss, J.: Implications of the observed anticorrelation between solar wind speed and coronal electron temperature. J. Geophys. Res. **108**, CideID 1158 (2003)
28. Fisk, L.A.: Acceleration of the solar wind as a result of the reconnection of open magnetic flux with coronal loops. J. Geophys. Res. **108**, 1157 (2003)
29. Bloch, T., Watt, C., Owens, M., et al.: Data-driven classification of coronal hole and streamer belt solar wind. Sol Phys **295**, 41 (2020). https://doi.org/10.1007/s11207-020-01609-z
30. Bravo, S., Stewart, G.A.: Fast and slow wind from solar coronal holes. Astrophys. J. **489**, 992 (1997)
31. Carpenter, D., Zhao, L., Lepri, S.T., Han, H.: Characterizing in-situ solar wind observations using clustering methods. In: Han, H., Baker, E. (eds.) SDSC 2022. CCIS, vol. 1725, pp. 125–138. Springer, Cham (2022). https://doi.org/10.1007/978-3-031-23387-6_9
32. Cranmer, S.R.: Coronal holes and the high-speed solar wind. Space Sci. Rev. **101**, 229 (2002)
33. Crooker, N.U., Antiochos, S.K., Zhao, X., Neugebauer, M.: Global network of slow solar wind. J. Geophys. Res. **117**, A04104 (2012). https://doi.org/10.1029/2011JA017236
34. Garrard, T., Davis, A., Hammond, J., et al.: The ACE science center. Space Sci. Rev. **86**, 649–663 (1998)
35. Gibson, S.E., Fan, Y.: The partial expulsion of a magnetic flux rope. Astrophys. J. **637**(1), L65–L68 (2006)
36. Gibson, S.E., Fan, Y., Török, T., Kliem, B.: The evolving sigmoid: evidence for magnetic flux ropes in the corona. Before During After CMEs Space Sci. Rev. **124**(1–4), 131–144 (2006)
37. Gloeckler, G., Cain, J., Ipavich, F.M., et al.: Investigation of the composition of solar and interstellar matter using solar wind and pickup ion measurements with SWICS and SWIMS on the ACE spacecraft. Space Sci. Rev. **86**, 497 (1998)
38. Han, H., Wentian, L., Wang, J., Qin, G., Qin, X.: Enhance explainability of manifold learning. Neurocomputing **500**, 877–895 (2022)
39. Crooker, N.U., McPherron, R.L.: Coincidence of composition and speed boundaries of the slow solar wind. J. Geophys. Res. **117**, A09104 (2012). https://doi.org/10.1029/2012JA017837
40. Ko, Y.-K., Raymond, J.C., Zurbuchen, T.H., et al.: Abundance variation at the vicinity of an active region and the coronal origin of the slow solar wind. Astrophys. J. **646**, 1275 (2006)
41. Lepri, S.T., Zurbuchen, T.H.: Iron charge state distributions as an indicator of hot ICMEs: possible sources and temporal and spatial variations during solar maximum. J. Geophys. Res. **109**, A01112 (2004). https://doi.org/10.1029/2003JA009954
42. Liu, S., Su, J.T.: Multi-channel observations of plasma outflows and the associated small-scale magnetic field cancellations on the edges of an active region. Astrophys. Space Sci. **351**, 417 (2014)
43. McComas, D., Bame, S., Barker, P., et al.: Solar wind electron proton alpha monitor (SWEPAM) for the advanced composition explorer. Space Sci. Rev. **86**, 563–612 (1998)
44. Aaron Roberts, D., et al.: Objectively determining states of the solar wind using machine learning. ApJ **889**, 153 (2020)
45. Smith, C., L'Heureux, J., Ness, N., et al.: The ACE magnetic fields experiment. Space Sci. Rev. **86**, 613–632 (1998)

46. Wang, Y.-M., Ko, Y.-K.: Observations of slow solar wind from equatorial coronal holes. Apj **880**, 146 (2019)
47. Wang, Y.-M., Grappin, R., Robbrecht, E., et al.: On the nature of the solar wind from coronal pseudostreamers. Astrophys. J. **749**, 182 (2012)
48. Wang, Y.-M., Ko, Y.-K., Grappin, R.: Slow solar wind from open regions with strong low-coronal heating. Astrophys. J. **691**, 760 (2009)
49. Zhao, L., et al.: An anomalous composition in slow solar wind as a signature of magnetic reconnection in its source region. ApJS **228**, 1 (2017)
50. Zirker, J.B.: Coronal holes and high-speed wind streams. Rev. Geophys. Space Phys. **15**, 257 (1977)
51. Zurbuchen, T.H., Fisk, L.A., Gloeckler, G., von Steiger, R.: The solar wind composition throughout the solar cycle: a continuum of dynamic states. Geophys. Res. Lett. **29**(9) (2002). https://doi.org/10.1029/2001GL013946
52. Han, H., Li, D., Liu, W., Zhang, H., Wang, J.: High dimensional mislabeled learning. Neurocomputing **573**, 127218 (2024)
53. Richardson, I.G., Cane, H.V.: Identification of interplanetary coronal mass ejections at 1 au using multiple solar wind plasma composition anomalies. J. Geophy. Res. Space Phys. **109** (2004). https://doi.org/10.1029/2004JA010598
54. Lepri, S.T., Zurbuchen, T.H., Fisk, L.A., Richardson I.G., Cane, H.V., Gloeckler, G.: Iron charge distribution as an identifier of interplanetary coronal mass ejections, J. Geophys. Res. **106**(A12), 29,231–29,238 (2001)

Node Classification with Multi-hop Graph Convolutional Network

Tonni Das Jui[1]([envelope])[iD], Mary Lauren Benton[1][iD], and Erich Baker[2]

[1] Baylor University, Waco, TX 76706, USA
tonni_jui1@baylor.edu
[2] Belmont University, Nashville, TN 37212, USA

Abstract. Various graph neural network architectures have emerged, each focusing on different facets of graphs, such as ensuring scalability, preserving local structures, and retaining structural and feature information within the node vectors. However, incorporating global structures into the latent representation is frequently overlooked despite being a central focus of many random-walk-based embedding techniques predating the advent of graph neural networks. Ensuring the preservation of global structural information enriches the node representations to tackle the node classification task effectively. We introduce an innovative multi-hopped graph convolutional network designed for graph-structured data to address this gap. This architecture leverages adjacent node information and incorporates distanced neighborhood information to preserve global structure in the latent representations. We propose a method for aggregating multiple hopped features and devise a straightforward yet effective architecture. Empirical validation of our theoretical findings on various graph classification benchmarks showcases that our model attains state-of-the-art and consistent performance.

Keywords: Graph embedding · Graph structure · Global information · Multi-hop · Node classification

1 Introduction

Graph-structured data, representing entities and their relationships, are ubiquitous across diverse domains ranging from social networks and biology to finance and recommendation systems. Due the additional benefits inherent in graph-structured data, such as their extensive relational information, there has been an increasing interest in collecting and integrating graph-structured data [9,20,22,27,31,36]. Unlocking the potential of such data requires sophisticated techniques capable of capturing both local and global structures [1,13,26]. In the era of big data, where complex relationships and intricate patterns are hidden in large networks, it is essential to leverage the latent knowledge encoded in graphs. However, the complexity of large graph structures makes identifying and using this latent knowledge challenging.

© The Author(s), under exclusive license to Springer Nature Switzerland AG 2024
H. Han (Ed.): SDSC 2024, CCIS 2158, pp. 199–213, 2024.
https://doi.org/10.1007/978-3-031-67871-4_14

Graph Neural Networks (GNNs), a revolutionary paradigm in machine learning, offer node embedding, a promising avenue for exploring the intricate nature of graph-structured data [20,32]. Node embedding generates latent representations or compressed node vectors, which we can use as the input vector for downstream graph tasks, such as graph clustering and node classification [11,32,34]. GCN's groundbreaking success in node classification set the stage for further GNN-based research on embeddings [20]. Unlike conventional feature engineering or matrix factorization techniques [8,9], GNN-based embeddings leverage the power of neural networks to learn representations of nodes and edges dynamically. These embedding approaches have gained popularity due to many successful implementations of neural network-based graph embedding techniques for node classification, node clustering, node attribute prediction, and graph visualization.

Current state-of-the-art GNN architectures focus either on a spectral convolution strategy comprising an adjacency matrix or by attending or inductively aggregating immediate (1-hop) neighbors. Through self-attention mechanisms, aggregation strategies, and sophisticated optimization, GNNs unravel the hidden semantics of nodes within a graph [14,20,32]. Hidden semantics of a graph refer to the implicit relationships, patterns, and associations between nodes and edges in the graph, which may not be apparent from the raw data but can be inferred through analysis or modeling techniques. However, the preservation of global structural information is often neglected by mainstream GNN models. Previous work has demonstrated the efficacy of leveraging multi-hop adjacency information in other embedding architectures for graph property identification, isomorphism testing, and distant neighbor identification.

Consistency in the performance scores has also become a focus for fairness research in machine learning that could be solved using latent representations [2,7,16,17]. Although a definite cause has yet to be discovered, recent work has suggested that bias (measurement bias, aggregation bias, label bias, etc.), model complexity, and models' sensitivity to limited data could be responsible for prediction inconsistency [5]. This inconsistency can make machine learning models unreliable and untrustworthy, leading to severe implications such as reinforcing discriminatory practices in the criminal justice system [18]. A resilient node vector representation, enriched with comprehensive global information, is instrumental in capturing important graph properties, thereby fostering consistency in the model and ensuring fairness.

Here, we develop a cutting-edge GNN-based approach that extends traditional methods in graph representation. We embed multiple hop information into node embeddings and analyze the experimental results comparing state-of-the-art GNN architectures. We also discuss the theoretical underpinnings, methodological intricacies, and practical implications of employing a multiple-hop-based GNN embedding approach. We summarize our contributions below:

- We propose and develop a simple but effective multi-hop convolutional propagation rule for GNN models where we utilize the weights of the connection between distanced neighbors (not just 1-hopped neighbors) and their features.

– We illustrate how the multi-hopped-based neural network architecture enhances the quality of embeddings and can be applied to node classification.
– We demonstrate that our method performs consistently on node classification tasks.

Our overall goal is to determine the impact of GNN-based embeddings on various downstream tasks, from node classification and link prediction to graph clustering and visualization. The rest of the paper is organized as follows. Section 2 discusses the background and approaches related to this method. We explain our architecture in Sect. 3 and discuss our experiments using the architecture in Sect. 4. We summarize our contributions and future research directions in Sect. 6.

2 Related Work

Many graph embedding methods have been proposed in the literature, each offering unique insights and capabilities for learning representations of nodes and edges in graphs. Among these, one category of graph embedding techniques leverages random walk-based approaches to capture local graph structures. Notable methods in this category include DeepWalk, which employs random walks to generate node sequences and then learns embeddings using skip-gram models [27]. Another random-walk-based method is Node2Vec, which introduces a flexible biased random walk strategy to explore the breadth and depth-first search patterns [13]. It extends the concept of word embeddings from natural language processing to graphs. This strategy allows it to capture both local and global graph structures effectively. By controlling two parameters, namely the return parameter p and the in-out parameter q, Node2Vec can bias the random walks to explore different neighborhoods of the graph. This approach presents a viable alternative for executing graph-related tasks relying on structural embeddings derived from graph data even in the era of GNN-based embeddings [15].

An alternative avenue of research centers on factorization-driven methodologies, which decompose the adjacency matrix of a graph to derive lower-dimensional representations. GraRep exemplifies this approach, which extends classical matrix factorization methodologies to accommodate higher-order proximities within graphs [8]. It addresses the inherent complexity of graph data by decomposing the adjacency matrix into lower-dimensional representations, facilitating more efficient processing and analysis. Similarly, HOPE utilizes matrix factorization coupled with asymmetric normalization techniques to capture both local and global structural characteristics [25]. The HOPE method emphasizes preserving higher-order proximity in the embedding. Although these methods effectively capture certain aspects of graph structure, they often require extensive computational resources and suffer from scalability issues, limiting their applicability to real-world datasets.

Moreover, graph embedding has seen significant advancements with the emergence of graph neural networks (GNNs), which have demonstrated superior performance in capturing intricate graph properties and learning representations from raw graph data. As a result, recent research prioritizes GNN-based approaches over traditional matrix factorization methods for their ability to handle diverse graph structures and tasks effectively. Methods such as Graph Convolutional Networks (GCNs) utilize convolutional operations over the graph to aggregate information from 1-hopped neighboring nodes and learn node representations [6, 10, 20].

Variants of GCNs, such as GraphSAGE, extend this framework to handle inductive learning tasks by recursive sampling and aggregating features from the neighborhood of each node [14]. This approach allows GraphSAGE to generalize well to unseen nodes and graphs, making it suitable for real-world applications where the graph structure may evolve or only partial information is available during training. Additionally, GraphSAGE offers flexibility in selecting aggregation functions and sampling strategies, enabling researchers to tailor the model to different graph domains and learning objectives. Furthermore, generative adversarial networks (GANs) have been explored for graph embedding tasks [33]. Adversarially Regularized Graph Autoencoders (ARGA) combine graph autoencoder architectures with adversarial training to learn embeddings that preserve local and global graph structures while generating realistic graph samples [26].

Other methods have also investigated graph dimension reduction techniques and the application of graph embedding techniques in multiple domains, including social networks [27], biology [13], recommendation systems [20], and knowledge graphs [4, 26]. These studies demonstrate the versatility and effectiveness of graph embedding techniques across diverse applications.

Despite the progress in this field, significant challenges remain, including scalability to large-scale graphs, capturing global information for node structures, and addressing issues of fairness and interoperability [21, 23, 30]. Addressing these challenges will be crucial for advancing the state-of-the-art in graph embedding research and unlocking the full potential of these techniques for real-world applications. Thus, our work focuses on preserving multiple-hopped information into node embeddings to gain viable global node information for node classification.

3   Multi-hop Architecture

This section provides theoretical motivation for a graph embedding representation that captures the global view of neighboring structures and explains multi-hop architecture.

3.1   Notations and Initial Setups

A graph $\mathcal{G}$ has a set of edges E in the form of adjacency matrix A, a set of nodes V, and node features X. Here, set $V = \{v_1, v_2, ...v_n\}$ consists of n nodes where the A will have a size of $n \times n$. If $(v_j, v_k) \in E$ then, $A_{j,k} = w_{v_j v_k}$ where $w_{v_j v_k} \neq 0$ is the associated weight of (v_j, v_k), otherwise $A_{j,k} = 0$. For unweighted graphs, $w_{v_j v_k} \in \{0, 1\}$ and for weighted graphs, $w_{v_j v_k}$ can be 0 and any number ≥ 1. The node features $X = \{f_{v_1}, f_{v_2}, ...f_{v_n}\}$ denotes feature vectors for n number of nodes. For a directed graph, $w_{v_j v_k} \neq w_{v_k v_j}$ and otherwise for undirected graphs. Our goal is to generate effective node embeddings $\mathbf{z} \in \mathbb{R}^{n \times d}$ for node classification.

3.2   Intuition

Before the emergence of neural-network-based graph embedding techniques (GraphSAGE [14], GCN [20], GAT [32], MPNN [11], GIN [34] etc.), random-walk-based approaches focused on capturing global structures by randomly traversing through multiple neighbors, as illustrated in Fig. 1 [13,27,31]. These models prepare multiple neighborhood sets from the random walk and define probabilistic approaches to preserve the neighborhood structure in the embedding. These models are known for preserving global structure and maintaining scalability and performance consistency. Randomly traversing neighboring nodes and integrating this information into the embedding process facilitates the preservation of global information within the dataset.

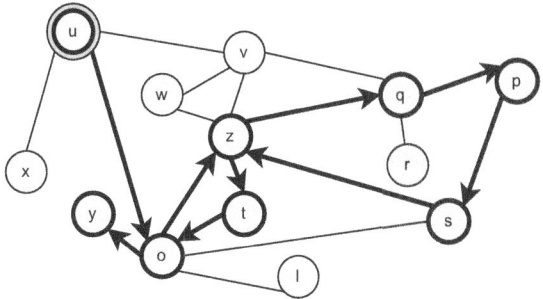

Fig. 1. Random walk from the source node A with a walk length of 10 consists of the path $u -> o -> z -> q -> p -> s -> z -> t -> o -> y$. The path is pointed through directed arrows. $N_{\text{Rand}_{10}} = \{u, o, z, q, p, s, t, y\}$ is the neighborhood set built from this random walk.

Alternatively, most neural-network-based graph embedding techniques developed node embeddings using direct neighborhood information as incorporating neighboring node information through random-walks into the neural network k architecture escalates model complexity. Direct neighborhood information or local neighboring represents information about neighbors until 1-hop. To capture global neighboring information into the embedding, we focus on incorporating multi-hopped neighborhood information into the convolution layer instead of randomly walked neighboring information. Both random walk techniques and multi-hopped techniques incorporates global information; however, random-walks escalates complexity while multi-hopping operation requires constant complexity. Concurrently, our objective entails crafting a GNN architecture distinct from random-walk-based methodologies, ensuring that the multi-hop node aggregation approach does not excessively increase model complexity.

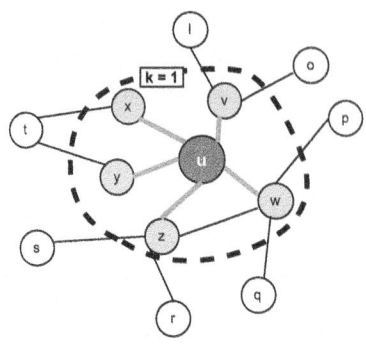

Fig. 2. 1-hop neighbors of node u are v, w, x, y, z nodes (when k=1). Popular graph embedding techniques preserve 1-hop neighbors feature and structure information into the embedding.

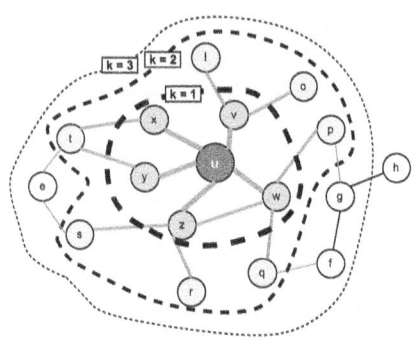

Fig. 3. Beyond the 1-hop neighbors, t, s, r, q, p, o and l are 2-hop neighbors of node u, where e, f, g are 3-hop neighbors.

Intuitively, nodes that possess information about a significant portion of their neighbors are more likely to capture global structural characteristics. Consequently, distant nodes are more likely to be classified into the same category when they possess information about the extent of their connectivity. We focus on incorporating multi-hop neighborhood information into the latent representation to encompass both immediate and distant neighbors. To construct the latent representation for a given node u, our approach involves not only attending to the 1-hop neighboring nodes x, y, z, w, and v as illustrated in Fig. 2, but also considering nodes with connections spanning multiple hops as depicted in Fig. 3.

3.3   Layer-Wise Transformation Rule of the Network

With this objective in mind, we establish a connection between distanced nodes u and v, denoted by $\tilde{A}_{uv}$ in Eq. (1), to leverage this connection in the embedding process.

$$\tilde{A}_{uv} = f(u, v)$$
$$\tilde{A}_{uv} = max(A_{uv}, \lambda_{uv}) + I_n \tag{1}$$

Here, the equation consists of adjacency/connection A_{uv}, relationship importance λ_{uv} and identity matrix I_n. Firstly, A_{uv} represents the connection between any two nodes u and v. In the original adjacency A, if u and v are connected by an edge, $A_{uv} = 1$, otherwise, $A_{uv} = 0$. We empirically formulated the function $f(u, v)$ to quantify the weight of the connection between any pair of nodes, indicating the level of attention required during embedding development. Equation (1) incorporates an identity matrix of size $n \times n$, I_n, into the adjusted adjacency information. As the modified adjacency matrix $\tilde{A}$ will be multiplied with the degree matrix $\tilde{D}$, it will enable the features of the neighboring nodes. However, we also require the embedding to incorporate the node's feature. Thus, we incorporate I_n into the modified adjacency matrix $\tilde{A}$ so that the later matrix multiplication ensures incorporating the node's own features as well as the neighboring nodes' features in the layer-wise transformation [20]. This is why the presence of I_n is crucial, as it ensures that the embedding network considers the feature information of the source node. Secondly, the formulation of the equation hinges upon the definition of λ_{uv} as well, which determines the relationship importance between nodes u and v.

$$\lambda_{uv} = k^{-1}.(k + 1 - d_{uv}) \qquad (2)$$

In this context, k represents the hop information, and through empirical analysis, we have determined that values ranging from 2 to 3 yield optimal performance for the citation networks under consideration [15]. To ensure regularization of the overall value, we introduce a regularization factor of k^{-1} into λ_{uv}. Additionally, d_{uv} represents the distance or edge count between nodes u and v. Leveraging $\tilde{A}$, we apply a straightforward two-layer convolution based on the convolutional layer-wise transformation principle outlined in GCN [20].

$$H^l = \sigma(\tilde{D}^{-\frac{1}{2}} \tilde{A} \tilde{D}^{-\frac{1}{2}} H^{l-1} W) \qquad (3)$$

The layer-wise transformation principle involves the degree matrix $\tilde{D} = D + I_n$ (for the same reason as mentioned above for Eq. (1)), where H^0 represents the feature matrix X. Activation functions such as ReLU and linear functions are applied to the layers specified in Eq. (4). This embedding methodology effectively encodes both the graph's structure and the content of its nodes, per the layer-wise transformation principle.

$$\begin{aligned} Z^1 &= f_{\text{Activation}}(X, A | W^{(0)}) \\ Z^2 &= f_{\text{Activation}}(Z^1, A | W^{(1)}) \end{aligned} \qquad (4)$$

The Eq. (4) stands for a two-layer graph convolutional network for node classification on a graph that utilizes the normalized adjacency information with multi-hop weights, $\tilde{A}$ and feature information X. W is the weight matrix that the neural network will learn.

Here, the Algorithm 1 illustrates a sequential overflow of our proposed method. Line 1 to line 5 depicts the modification of adjacency information according to relationship strength. Line 7 to 9 illustrates the layer-wise propagation. As we implement GCN's convolution filter, our model's convolution is polynomial in the Laplacian, resulting in $\mathcal{O}(|E|)$, according to the number of edges [20]. The incorporation of k-hopped information is a matrix multiplication operation in parallel to other matrix multiplications inside the layer-wise propagation. Thus, our model's layer-wise propagation has the same complexity as GCN's declared complexity, $\mathcal{O}(|E|FC)$, where E and F are number of edges and features.

Algorithm 1. Multi-hop graph convolution

Require: $G = \{A, V, X\}$ where A, V, X are the adjacency matrix, set of nodes, and the feature matrix accordingly; n= number of nodes; m= number of node pairs/edges; L= number of layers.

1: **for** $u = 1, 2, 3, \ldots, n$ **do,**
2:     **for** $v = 1, 2, 3, \ldots, n$ **do,**
3:        Develop relationship importance λ_{uv} according to Eq. (2)
4:        Generate $\tilde{A}_{uv}$ utilizing λ_{uv} following Eq. (1)
5:     **end for**
6: **end for**
7: **for** $l = 1, 2, 3, \ldots, L$ **do,**
8:     Regularize $\tilde{A}$ by row normalization with $\tilde{D}$ and develop output H^l from the output of previous layers, H^{l-1} following Eq. (3)
9: **end for**
10: Output $H^L \in \mathbb{R}^{n \times d}$ as the final node representations

---

4 Experiments

In our exploration of node classification, we closely follow the experimental guidelines established by Yang et al. [35], a methodology broadly embraced in current graph embedding studies [8, 14, 20, 26].

4.1 Networks

We employ citation networks, including Citeseer, Cora, and Pubmed [29], with dataset particulars outlined in Table 1. In these networks, nodes represent documents and have bag-of-words feature vectors, while edges signify citation links (undirected).

Table 1. Dataset statistics.

Dataset	Graph Type	Nodes	Edges	Features	Node classes
Citseer	Citation network	3327	4732	3703	6
Cora	Citation network	2708	5429	1433	7
Pubmed	Citation network	19717	44338	5414	210

4.2 Setups

In our training regimen, we adhere to the protocol outlined by GCN [20]. Using the same dataset partitions as described by Yang et al. [35], we assess predictive accuracy on a test subset comprising 1000 labeled instances, employing the two-layer convolutional network detailed in Eq. (4). Hyperparameter optimization is performed on a validation set containing 500 labeled instances, although validation labels are not employed during training. Key hyperparameters include the dropout rate for all layers, the L2 regularization factor for the initial convolutional layer, and the number of hidden units. Training iterations are capped at 200, with the Adam optimizer [19] employed, featuring a learning rate of 0.01 and early stopping criteria set at a window size of ten epochs (terminating training if validation loss stagnates for ten consecutive epochs). Weight initialization follows Glorot's method, incorporating (row-)normalization for input feature vectors [12].

4.3 Baselines

In the process of selecting baseline methods for comparative analysis with state-of-the-art approaches, we reference GAT [32], GIN [34], MoNet [24], and GNN [28], identifying common choices across these sources. Our chosen baselines include GCN [20] and GraphSAGE [14], suitable graph embedding methodologies focusing on node classification tasks. GCN performs convolutional operations in the graph domain, similar to convolutional neural networks (CNNs) in the image domain [3,4]. It normalizes the adjacency and feature matrix multiplication into the convolutional layer. GraphSAGE works inductively, leveraging multiple layers of aggregators to capture different levels of neighborhood information [14]. Additionally, we incorporate Node2Vec [13] and Attri2Vec [36] into our selection criteria. Node2Vec, leveraging a probabilistic framework, effectively embeds structural insights into latent representations by employing random-walk strategies, thereby capturing global structural characteristics within each node. Conversely, Attri2Vec, which integrates feature vectors into embeddings and leverages a feature encoder, prioritizes neighborhood preservation and attributes information to node embeddings [36]. Given the alignment between the architectural objectives of Node2Vec, Attri2Vec, and their targeted graph task with our proposed method, we find it pertinent to incorporate Attri2Vec. We compare our results against these four methods to validate the efficacy of our approach.

4.4 Metrics

We adhere to the performance evaluation standards in the referenced literature, and thus, we employ the classification accuracy [14,20,35,36]. We also consider the F1_macro metric. While alternative metrics such as accuracy or weighted F1 scores are available, F1_macro has been deemed suitable for ensuring balanced performance in multi-class classification scenarios, aligning with the characteristics of our dataset [1]. F1_macro computes individual scores for each class and subsequently averages them, providing equal importance to all classes. In contrast, accuracy quantifies the proportion of correctly classified instances, whereas F1_macro assesses the model's overall effectiveness by considering precision and recall across all classes equally, thus mitigating the impact of class imbalance.

5   Results

Our study assesses the model's representational capacity by comparing its testing accuracy to other state-of-the-art methods on three datasets. Our findings, outlined in Table 2, portray the accuracy of node classification expressed as a percentage. To ensure robustness and reliability in our evaluations, we conduct 10 independent runs with random node orderings. We obtained the F1_macro scores by aggregating the outcomes from a single experimental run. Each cell in Table 2 represents a value denoted as $a \pm b$ where the mean accuracy (a) is accompanied by the standard error (b). The standard error measures the precision with which a sample estimates the population. Conceptually, it quantifies the extent to which the sample mean deviates from the true population mean, offering insights into the reliability of our findings. Our observed scores for GCN align closely with those reported in the seminal GCN paper [20], validating the fidelity of our experimental setup. Moreover, we meticulously replicated the experimental conditions of other baseline models, ensuring consistency and comparability across our evaluations.

Table 2. Classification accuracies (%) for test sets of the datasets through the first eight rows and then the F1_macro scores through the rest.

	Method	Cora	Citeseer	Pubmed
Accuracy	GCN	81.5 ± 0.008	70.3 ± 0.007	79.0 ± 0.009
	GraphSAGE	84.0 ± 0.012	83.0 ± 0.011	80.6 ± 0.015
	Node2Vec	79.8 ± 0.004	76.1 ± 0.005	80.7 ± 0.003
	Attri2Vec	78.3 ± 0.009	82.1 ± 0.008	79.6 ± 0.006
	ARGA	79.3 ± 0.009	77.6 ± 0.009	80.1 ± 0.011
	2-hop GCN	**86.8 ± 0.002**	**82.0 ± 0.001**	**79.8 ± 0.004**
	3-hop GCN	**85.9 ± 0.005**	**86.7 ± 0.004**	**83.3 ± 0.006**
	4-hop GCN	65.0 ± 0.006	80.9 ± 0.003	70.4 ± 0.003
F1_macro	GCN	0.29	0.28	0.27
	GraphSAGE	0.31	0.34	0.32
	Node2Vec	0.27	0.25	0.27
	Attri2Vec	0.27	0.26	0.29
	ARGA	0.21	0.23	0.24
	2-hop GCN	**0.31**	**0.35**	**0.35**
	3-hop GCN	**0.34**	**0.36**	**0.34**
	4-hop GCN	0.25	0.23	0.22

We hypothesize that the best training and testing set accuracy would come from models that could correctly include global information in the embedding. This eventually leads to their heightened representational capacity. The representational capacity in the context of graph embedding refers to the ability of a model or technique to effectively capture and encode the essential characteristics and information of a graph within its learned embeddings. It reflects how well the embeddings represent the structural and semantic properties of the nodes and edges in the graph. A higher representational capacity implies that the embeddings can accurately reflect the intricate relationships, patterns, and features in the graph, enabling better performance in downstream tasks such as node classification, link prediction, or graph clustering. We anticipate that the multi-hop convolutional method, designed to fit extensive neighborhood structural insights, would have enhanced node representational power and yield a discernible performance increase relative to other methods.

GraphSAGE, Attri2Vec, and ARGA claim to be focused on global information preservation. Although most of these approaches score well for all three datasets, as shown in Table 2, GraphSAGE, Attri2Vec, and ARGA have a higher standard deviation than GCN and Node2Vec. On the other hand, the k-hopped convolutional network performs consistently in node classification tasks (for $k = 2$ and 3) as it has a comparatively lower standard error. The variability in standard error can be attributed to the sensitivity of models to data and parameter tuning. In small datasets like Cora, Citeseer, and Pubmed, where labels are not available for all nodes, training, and testing set sizes are usually constrained, which may result in significant variation in size, leading to diverse outcomes across experiment runs and, consequently, a more significant standard error. Additionally, variations in hyperparameter settings and the complexity of models, such as the GraphSAGE model with intricate *aggregate* and *combine* functions or techniques like Attri2vec and ARGA based on adversarial autoencoders, can contribute to differing classification results and increased standard error. Comparatively, simpler architectures like GCN and techniques like Node2vec exhibit lower standard errors due to their architectural simplicity. Our model integrates insights from both GCN and Node2vec to enhance the expressiveness of generated embeddings, resulting in improved performance and a reduced standard error, highlighting its effectiveness.

We observe that k-hopped convolution fails to produce expected accuracy where $k \geq 4$ for our baseline datasets, which aligns with the insights from existing literature of Hamilton et al. [14] and Tonni et al. [15]. Hamilton et al. suggested aggregating features from 1-hopped neighbors recursively until three iterations for these same citation datasets, as further iterations will only concatenate the same feature vectors instead of contributing new information to the node representations [14]. For predicting k-distanced neighbors of a node within the graph, Tonni et al. proposed confining the prediction scope to neighbors at a distance of 2–3, as extending beyond this range would excessively increase computational overhead and potentially diminish performance outcomes [15]. These insights explain why a convolutional network consisting of 4 or more connection weights degrades the performance and becomes inconsistent. Figure 4, Fig. 5, and Fig. 6 show the low standard errors for k-hopped approaches, further corroborating that the k-hopped convolutional network performs well. In all these figures, k-hop GCNs have lower standard errors than others.

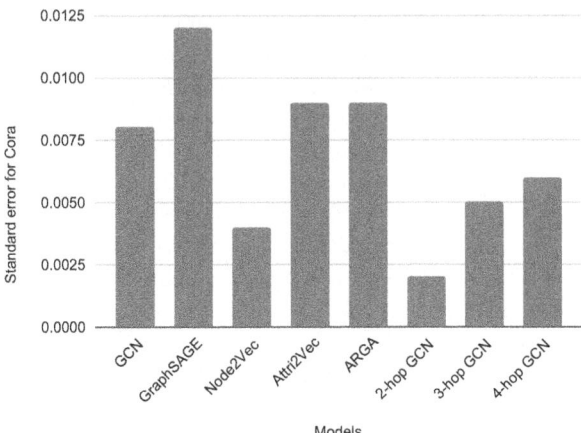

Fig. 4. A comparison of standard errors derived from ten iterations across various approaches for node classification task accuracy scores is depicted (for the Cora dataset). A shorter bar corresponds to a superior performance of the approach in maintaining consistency across scores. The visual representation highlights the 2-hop convolution as the most stable method for the dataset under consideration.

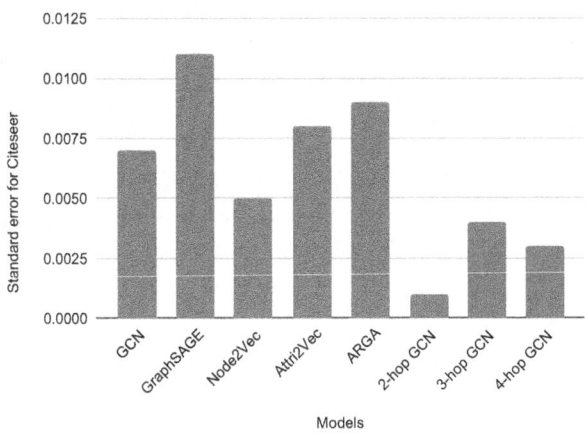

Fig. 5. A comparison of standard errors derived from ten iterations across various approaches for node classification task accuracy scores is depicted (for the Citeseer dataset). A shorter bar corresponds to a superior performance of the approach in maintaining consistency across scores. Here, 2-hop convolution maintains the most stable performance.

Moreover, the k-hopped GCN outperforms the other approaches in terms of both accuracy and F1_macro scores. As F1_macro computes individual scores for each class and subsequently averages them, providing a balanced performance score throughout all classes, the F1_macro scores are much smaller than the accuracy scores. Thus, utilizing F1_macro to measure performance ensures an unbiased embedding model.

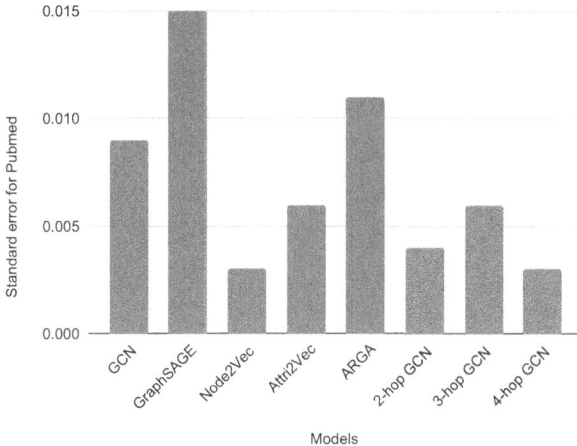

Fig. 6. A comparison of standard errors derived from ten iterations across various approaches for node classification task accuracy scores is depicted (for the Pubmed dataset). A shorter bar corresponds to a superior performance of the approach in maintaining consistency across scores. Here, 2-hop convolution maintains the most stable performance.

6   Conclusion

In conclusion, we propose an embedding approach that calculates connection weights throughout distanced neighbors and utilizes these weights and node features into a convolutional layer-wise transformation (GCN) to capture a global structural view of the graph. Our proposed method successfully encompasses global structural and feature information into the node embedding. The developed method also showcases consistent performance on accuracy scores for node classification tasks on our baseline datasets, which ensures the model is free of inconsistency bias. We analyze the performance of the proposed method against four baseline models that aim to capture a global view of the graph in various strategies and show how our method outperforms the performance of the existing approaches.

References

1. Ali, S., Shakeel, M.H., Khan, I., Faizullah, S., Khan, M.A.: Predicting attributes of nodes using network structure. ACM Trans. Intell. Syst. Technol. (TIST) **12**(2), 1–23 (2021)
2. Bellamy, R.K.E., et al.: AI Fairness 360: an extensible toolkit for detecting, understanding, and mitigating unwanted algorithmic bias. arXiv:1810.01943 [cs] (2018)
3. Biswas, D., Rahman, M.M.M., Zong, Z., Tešić, J.: Improving the energy efficiency of real-time DNN object detection via compression, transfer learning, and scale prediction. In: 2022 IEEE International Conference on Networking, Architecture and Storage (NAS), pp. 1–8 (2022). https://doi.org/10.1109/NAS55553.2022.9925528

4. Biswas, D., Tešić, J.: Small object difficulty (SOD) modeling for objects detection in satellite images. In: 2022 14th International Conference on Computational Intelligence and Communication Networks (CICN), pp. 125–130 (2022). https://doi.org/10.1109/CICN56167.2022.10008383

5. Black, E., Fredrikson, M.: Leave-one-out unfairness. In: Proceedings of the 2021 ACM Conference on Fairness, Accountability, and Transparency, pp. 285–295 (2021). https://doi.org/10.1145/3442188.3445894

6. Bruna, J., Zaremba, W., Szlam, A., LeCun, Y.: Spectral networks and locally connected networks on graphs. arXiv preprint arXiv:1312.6203 (2013)

7. Calders, T., Žliobaitė, I.: Why unbiased computational processes can lead to discriminative decision procedures. In: Custers, B., Calders, T., Schermer, B., Zarsky, T. (eds.) Discrimination and Privacy in the Information Society. Studies in Applied Philosophy, Epistemology and Rational Ethics, vol. 3, pp. 43–57. Springer, Heidelberg (2013). https://doi.org/10.1007/978-3-642-30487-3_3

8. Cao, S., Lu, W., Xu, Q.: GraRep: learning graph representations with global structural information. In: Proceedings of the 24th ACM International on Conference on Information and Knowledge Management, pp. 891–900 (2015)

9. Cao, S., Lu, W., Xu, Q.: Deep neural networks for learning graph representations. In: Proceedings of the AAAI Conference on Artificial Intelligence, vol. 30 (2016)

10. Defferrard, M., Bresson, X., Vandergheynst, P.: Convolutional neural networks on graphs with fast localized spectral filtering. In: Advances in Neural Information Processing Systems, vol. 29 (2016)

11. Gilmer, J., Schoenholz, S.S., Riley, P.F., Vinyals, O., Dahl, G.E.: Neural message passing for quantum chemistry. In: International Conference on Machine Learning, pp. 1263–1272. PMLR (2017)

12. Glorot, X., Bengio, Y.: Understanding the difficulty of training deep feedforward neural networks. In: Proceedings of the Thirteenth International Conference on Artificial Intelligence and Statistics, pp. 249–256. JMLR Workshop and Conference Proceedings (2010)

13. Grover, A., Leskovec, J.: node2vec: scalable feature learning for networks. In: Proceedings of the 22nd ACM SIGKDD International Conference on Knowledge Discovery and Data Mining, pp. 855–864 (2016)

14. Hamilton, W., Ying, Z., Leskovec, J.: Inductive representation learning on large graphs. In: Advances in Neural Information Processing Systems, vol. 30 (2017)

15. Jui, T.D., Baker, E., Benton, M.L.: k-hopped link prediction with graph embedding. In: 2023 International Conference on Computational Science and Computational Intelligence (CSCI), vol. 2023 (2023)

16. Jui, T.D., Rivas, P.: Fairness issues, current approaches, and challenges in machine learning models. Int. J. Mach. Learn. Cybern. 1–31 (2024)

17. Kamiran, F., Calders, T.: Classifying without discriminating. In: 2009 2nd International Conference on Computer, Control and Communication, pp. 1–6. IEEE (2009)

18. Kamiran, F., Karim, A., Zhang, X.: Decision theory for discrimination-aware classification. In: 2012 IEEE 12th International Conference on Data Mining, pp. 924–929. IEEE (2012)

19. Kinga, D., Adam, J.B., et al.: A method for stochastic optimization. In: International Conference on Learning Representations (ICLR), San Diego, California, vol. 5, p. 6 (2015)

20. Kipf, T.N., Welling, M.: Semi-supervised classification with graph convolutional networks. arXiv preprint arXiv:1609.02907 (2016)

21. Kleinberg, J., Mullainathan, S., Raghavan, M.: Inherent trade-offs in the fair determination of risk scores. arXiv preprint arXiv:1609.05807 (2016)
22. Meng, Z., Adluru, N., Kim, H.J., Fung, G., Singh, V.: Efficient relative attribute learning using graph neural networks. In: Proceedings of the European Conference on Computer Vision (ECCV), pp. 552–567 (2018)
23. Mishler, A., Kennedy, E.H., Chouldechova, A.: Fairness in risk assessment instruments: post-processing to achieve counterfactual equalized odds. In: Proceedings of the 2021 ACM Conference on Fairness, Accountability, and Transparency, FAccT 2021, pp. 386–400. Association for Computing Machinery, New York (2021). https://doi.org/10.1145/3442188.3445902
24. Monti, F., Boscaini, D., Masci, J., Rodola, E., Svoboda, J., Bronstein, M.M.: Geometric deep learning on graphs and manifolds using mixture model CNNs. In: Proceedings of the IEEE Conference on Computer Vision and Pattern Recognition, pp. 5115–5124 (2017)
25. Ou, M., Cui, P., Pei, J., Zhang, Z., Zhu, W.: Asymmetric transitivity preserving graph embedding. In: Proceedings of the 22nd ACM SIGKDD International Conference on Knowledge Discovery and Data Mining, pp. 1105–1114 (2016)
26. Pan, S., Hu, R., Long, G., Jiang, J., Yao, L., Zhang, C.: Adversarially regularized graph autoencoder for graph embedding. arXiv preprint arXiv:1802.04407 (2018)
27. Perozzi, B., Al-Rfou, R., Skiena, S.: DeepWalk: online learning of social representations. In: Proceedings of the 20th ACM SIGKDD International Conference on Knowledge Discovery and Data Mining, pp. 701–710 (2014)
28. Scarselli, F., Gori, M., Tsoi, A.C., Hagenbuchner, M., Monfardini, G.: The graph neural network model. IEEE Trans. Neural Netw. **20**(1), 61–80 (2008)
29. Sen, P., Namata, G., Bilgic, M., Getoor, L., Galligher, B., Eliassi-Rad, T.: Collective classification in network data. AI Mag. **29**(3), 93–93 (2008)
30. Stevenson, M.: Assessing risk assessment in action. Minn. L. Rev. **103**, 303 (2018)
31. Tang, J., Qu, M., Wang, M., Zhang, M., Yan, J., Mei, Q.: LINE: large-scale information network embedding. In: Proceedings of the 24th International Conference on World Wide Web, pp. 1067–1077 (2015)
32. Veličković, P., Cucurull, G., Casanova, A., Romero, A., Lio, P., Bengio, Y.: Graph attention networks. arXiv preprint arXiv:1710.10903 (2017)
33. Wang, D., Cui, P., Zhu, W.: Structural deep network embedding. In: Proceedings of the 22nd ACM SIGKDD International Conference on Knowledge Discovery and Data Mining, pp. 1225–1234 (2016)
34. Xu, K., Hu, W., Leskovec, J., Jegelka, S.: How powerful are graph neural networks? arXiv preprint arXiv:1810.00826 (2018)
35. Yang, Z., Cohen, W.W., Salakhutdinov, R.: Revisiting semi-supervised learning with graph embeddings. In: Proceedings of the 33rd International Conference on International Conference on Machine Learning, ICML 2016, vol. 48, pp. 40–48. JMLR.org (2016)
36. Zheng, P., Wen, Y., Chen, M., Chen, G.: attr2vec: learning node representations from attributes of nodes. In: Wu, Q., Zhao, K., Ding, X. (eds.) WiSATS 2020, Part II. LNICST, vol. 358, pp. 546–555. Springer, Cham (2021). https://doi.org/10.1007/978-3-030-69072-4_44

PyDaskShift: Automatically Convert Loop-Based Sequential Programs to Distributed Parallel Programs

Agm Islam and Greg Speegle[✉]

Baylor University, Waco, TX 76798, USA
{agm_islam1,greg_speegle}@baylor.edu

Abstract. In computing, parallel and distributed programming has become vital for processing large datasets and solving complex problems. Dask, an open-source library for Python, simplifies parallel computation by integrating seamlessly with Python's ecosystem. We introduce a new tool PyDaskShift to automatically transform sequential Python programs into their Dask Bag API equivalents. Currently, no automated tool performs this conversion, highlighting PyDaskShift's significance in advancing parallel programming automation in Python. The tool analyzes the program's structure, functions, and loops using Python's Abstract Syntax Tree (AST). By identifying opportunities for paralleliza-tion, it applies the concept of Gradual Synthesis for Static Parallelization to the identified code sections. After generating the Dask Bag program, PyDaskShift verifies its correctness using user-provided test cases. If any test case fails, the tool proceeds to the next phase of gradual synthe-sis. We evaluated PyDaskShift, and the resulting Dask Bag programs achieved performance gain in execution time up to 18 times.

Keywords: Program synthesis · distributed computing · Dask

1 Introduction

In the rapidly evolving landscape of computing, the pursuit of enhanced per-formance and computational efficiency has driven the advent of parallel pro-gramming [3]. Historically, computer systems were predominantly equipped with single-core processors that executed tasks sequentially. However, the demand for processing vast datasets and solving increasingly complex problems has necessi-tated a paradigm shift. Parallel programming is the answer, allowing developers to exploit the potential of multi-core processors, clusters, and multiple machines in data warehouses. It enables us to decompose intricate computational prob-lems into smaller, parallelizable tasks, which can be executed concurrently. This parallel execution not only enhances computational speed but also efficiently uti-lizes the available hardware resources, making it indispensable for meeting the demands of today's data-driven applications, scientific simulations, and high-performance computing [7].

H. Han (Ed.): SDSC 2024, CCIS 2158, pp. 214–234, 2024.
https://doi.org/10.1007/978-3-031-67871-4_15

While parallel programming offers immense benefits, it presents challenges, notably in coordination, shared resource management, and synchronization [13]. Developers must employ advanced techniques and tools to address these issues effectively. The transition from sequential programming to parallel programming can be complex and requires a deep understanding of parallel computing concepts [18]. Parallel programming poses a considerable challenge for developers, both in terms of acquiring the necessary expertise and achieving efficiency in its utilization. This complexity stems from the specialized knowledge it demands, making it a multifaceted process. Despite these inherent difficulties, there have been significant strides in the development of frameworks designed to streamline and simplify parallel programming [5]. These frameworks aim to provide a more accessible entry point for programmers looking to harness the power of parallelism. While these frameworks undoubtedly offer an advantage by abstracting some of the intricacies of parallel programming, they do not eliminate the need for developers to possess a profound and comprehensive understanding of how the frameworks operate. In essence, the frameworks act as a bridge to parallelism, making it more approachable, but they do not eliminate the need for a solid foundational knowledge of the underlying principles.

Dask [8] stands out as a noteworthy framework that leverages the native capabilities of Python, enabling developers to harness the power of parallel computation. This open-source library empowers users to perform parallel processing, making it an essential tool for tasks that require the concurrent execution of computational operations [21]. Dask is particularly attractive due to its seamless integration with the Python ecosystem. By virtue of its Python-native design, it aligns well with the existing Python data science and computational libraries. This compatibility facilitates the transition to parallel computing for Python developers but it requires mastering Dask programming.

PyDaskShift introduces a novel capability to transform sequential programs into their equivalent program using Dask Bag API. PyDaskShift employs classical compiler techniques such as the application of pattern matching rules [19]. This approach allows PyDaskShift to identify and convert specific code patterns within code fragments into the desired target framework. Specifically, it can convert a segment of code containing loops into a Dask bag program with similar functionality. To the best of our current knowledge, there exists no automated tool as of now that performs this conversion, highlighting the significance of PyDaskShift in advancing the automation of parallel programming in Python. PyDaskShift employs a substantial portion of the Dask Bag API, although it does not utilize its complete set of functionalities.

PyDaskShift is designed to accept a sequential Python program as its primary input, in addition to two important parameters: the block size and the number of workers. These parameters are vital for configuring the resulting Dask Bag program. The block size in Dask refers to the size of data chunks used when reading, allowing parallel operations on each chunk. Dask workers function as independent computing entities responsible for carrying out computations within a Dask cluster. They serve as the parallel processing components and collaborate to complete tasks and perform data operations. When PyDaskShift encounters file reading operations within the program, it relies on the block size to parallelize these read operations effectively. The number of workers, on the other hand, determines how many concurrent computations can be executed.

To achieve parallelization, PyDaskShift collects and stores information about the program structure, functions, and loops using Python's Abstract Syntax Tree (AST). By analyzing the stored data, the tool identifies opportunities where parallelization can be applied. PyDaskShift applies the concept of Gradual Synthesis for Static Parallelization [11] to the identified sections of code that need conversion.

Once the code generation phase is finished, the tool proceeds to verify the translated program's correctness using test cases supplied by the user. The comparison of results with these test cases aims to establish equivalence between the original and translated programs [15]. If any of the test cases fails, it indicates an unsuccessful program translation, prompting the tool to advance to the next stage of gradual synthesis.

The structure of this work can be outlined as follows: Sect. 2 provides essential background information and reviews related works in the field. Section 3 offers an in-depth exploration of the overall system architecture, providing explanations for each component. Section 4 validates the outcomes of PyDaskShift using a set of experiments and offers a detailed discussion of the results. Finally, Sect. 5 provides a summary and discusses potential avenues for future research and development.

2   Related Works

This project uses a technique called gradual synthesis to generate a Dask program from a sequential python program. It is similar to other techniques used by Casper [9] and Diablo [12]. It extends the work on Tyro [20]. In this section, we review these techniques.

2.1   Program Synthesis

Program synthesis refers to the automated process of generating computer programs from a given specification [17]. The paper [11] proposes an automated approach called Gradual Synthesis for Static Parallelization (GRASSP). GRASSP aims to find a new program that preserves equivalence with the serial one, addressing data dependencies and gradually developing solutions. It leverages automated formal methods and manages data approximations and dependencies, making its results adaptable to various programming platforms. PyDaskShift leverages gradual synthesis by systematically seeking the optimal solution for a function. It begins with most straightforward solutions and progressively increases complexity until it identifies the most suitable one.

2.2   Data-Intensive Array-Based Loop Optimizer (DIABLO)

In their research [12], the authors proposed a framework, Data-Intensive Array-Based Loop Optimizer (DIABLO), that can transform array-based programs into distributed data-parallel applications. Their conceptual model converts the input program into monoid comprehensions resembling SQL statements. The authors employed a key-value map to represent sparse arrays, vectors, or matrices. The framework simplifies the process by eliminating the iteration step in the first phase by identifying an appropriate inverse function, such that $f(F(K)) = k$, where k represents the index array type. Without such inversion, conversion is not feasible. They employ group-by and aggregation techniques to manage incremental changes, ultimately producing the converted monoid comprehension.

PyDaskShift is fundamentally different from DIABLO in that we use a compiler technique to identify the possible translations and use program testing to verify the results. This should enable us to identify more translations, but with the limitation that the verification may not be sufficient.

2.3   Casper

Casper [2] is designed to automate the conversion of sequential Java code into MapReduce programs suitable for execution in distributed computing frameworks like Hadoop [22], Spark [23], and Flink [16]. The CASPER tool encompasses a three-step process that motivated Tyro and thus PyDaskShift.

The initial step in CASPER's conversion process, the *progam analyzer*, parses the input Java program to create an Abstract Syntax Tree. From this tree, CASPER identifies code segments within the program that have the potential for translation into MapReduce constructs. CASPER encodes these selected code fragments into a high-level intermediate representation (IR).

CASPER uses the IR in a *summary generator* to synthesize and verify program summaries. These program summaries are verified against the program's postconditions. With the verified program summaries, CASPER executes its *code generator* to create executable MapReduce code. CASPER aims to specify reliable summaries by applying Hoare logic-based verification criteria [14], expressed as Boolean predicates.

CASPER also automates the generation of verification criteria and runtime checks for the programs. In cases where verification fails, CASPER provides a new grammar to the synthesizer to prevent the generation of the same summary. Since multiple MapReduce programs can be generated, CASPER employs a runtime monitoring module. This module evaluates the costs of different MapReduce programs using a sample input dataset. Once the evaluation is complete, CASPER generates the final equivalent MapReduce program that optimally suits the requirements.

PyDaskShift exhibits certain resemblances to Casper in terms of its functionality. For instance, PyDaskShift and Casper both rely on Abstract Syntax Trees (AST) for static code analysis, a common element between them. However, there is a notable distinction in how they handle code transformation. While Casper employs an intermediate language to facilitate the conversion of code fragments, PyDaskShift takes a different approach. Instead of utilizing an intermediate language, PyDaskShift harnesses metadata to guide and execute the translation of operations within the code. Furthermore, the verification process in PyDaskShift deviates significantly from that of Casper. In Casper, the verification process is an automated part of the tool, whereas in PyDaskShift, it relies on test cases supplied by the user.

2.4   Tyro

Tyro [20] is designed to transform a sequential program into a parallel program within the PySpark framework. The primary focus of their study was on optimizing *for* loops that were suitable for parallelization. Similar to Casper, Tyro is structured into three distinct phases to achieve this goal: program analysis; operation translation and code generation and verification.

In the *progam analysis* phase, Tyro utilizes the Python AST (Abstract Syntax Tree) module to extract metadata from the code. This metadata encompasses details related to the program, such as functions, and details related to the code snippet, such as loops, and operations. Next, in the *operation translation*, Tyro translates the code into the map, reduce, or a pattern-matching program based on synthesis parameter. The Synthesis parameter determines which operation to employ. Finally, the code is generated and verified. by using test cases provided by the user. If a test case fails, the software reverts back to the Operation Translation stage, implementing a more advanced equivalency. If no new equivalencies remain, the program terminates.

The goal of PyDaskShift is to enhance performance, matching the speed of Tyro using a pure python based distributed environment as the synthesis goal. The code produced by Tyro executes approximately 6.2 times faster than the original sequential program. PyDaskShift saw slightly better performance gains, with a speedup of 7 in a parallel environment. While its architectural design and workflow closely resemble those of Tyro, PyDaskShift distinguishes itself by supporting a different distributed programming system (Dask).

3   Methodology

In this section, we will provide a simple example to demonstrate the functionalities of PyDaskShift. While more complex cases require repeating these steps, this case serves as template for all compilation. The comprehensive implementation can be accessed from the Git repository provided in the references [1].

PyDaskShift accepts a conventional Python program that follows a sequential execution paradigm and generates an equivalent program utilizing the Dask Bag API. Its primary function is to facilitate the seamless transformation of the input program's logic into a parallelized Dask Bag implementation. The change of a sequential Python program into a Dask Bag program encompasses a series of stages that each play a role in this conversion process:

- Program Analyzer
- Feature Extractor
- Component Transformer
- Code Generator
- Verifier

Sanjel and Speegle, showcased the transformation of sequential Python programs into PySpark programs in Tyro [20]. Our work builds upon this foundation by adhering to their architecture while introducing modifications to the software and removing limitations. Figure 1 depicts the architecture of PyDaskShift. Due to space limitations, we only present the major changes from Tyro. The reader can refer to [20] for the missing details.

Figure 2 shows an example of a sequential Python program that iterates over a list of numbers and finds the minimum number. This program is the input to PyDaskShift. As in Tyro, the Python AST module scans the input program to create an AST. The Program Analyzer extracts comprehensive program information from the AST, referred to as meta-information. This meta-information is accessible to other PyDaskShift components. The Feature Extractor retrieves information related to functions, loops, and operations within loops from the meta-information provided by the Program Analyzer. Likewise, the program verification follows the Tyro model.

The primary contribution of this paper is the novel Component Transformer (Sect. 3.1) for PyDaskShift. This transformer utilizes the meta-information to translate operations inside loops into equivalent Dask Bag operations. The Bag operations are then forwarded to the Code Generator (Sect. 3.8), which replaces sequential operations with the Dask Bag operations, includes necessary imports, addition of global variables, and generates the complete code.

At this stage, we are unable to provide an assurance regarding the accuracy of the program. The Verifier module assumes the responsibility of running the generated code on the provided test cases. Should the test cases successfully validate the generated code, it will then be designated as the definitive output of PyDaskShift. If the tests do not pass successfully, the synthesis stage advances to the subsequent level, commencing anew with the Component Transformer.

3.1   Component Transformer

The Component Transformer initiates its process with meta-information, such as the identification of files, loops and other operations. During this stage, the module converts the intermediate form of the extracted information into the equivalent Dask Bag program, driven by the synthesis stage. It commences with static operations to convert files and other structures, then proceeds with the map operations and, in the event of verification failures, proceeds to more complex operations as defined by the synthesis stage. PyDaskShift encompasses various synthesis stages:

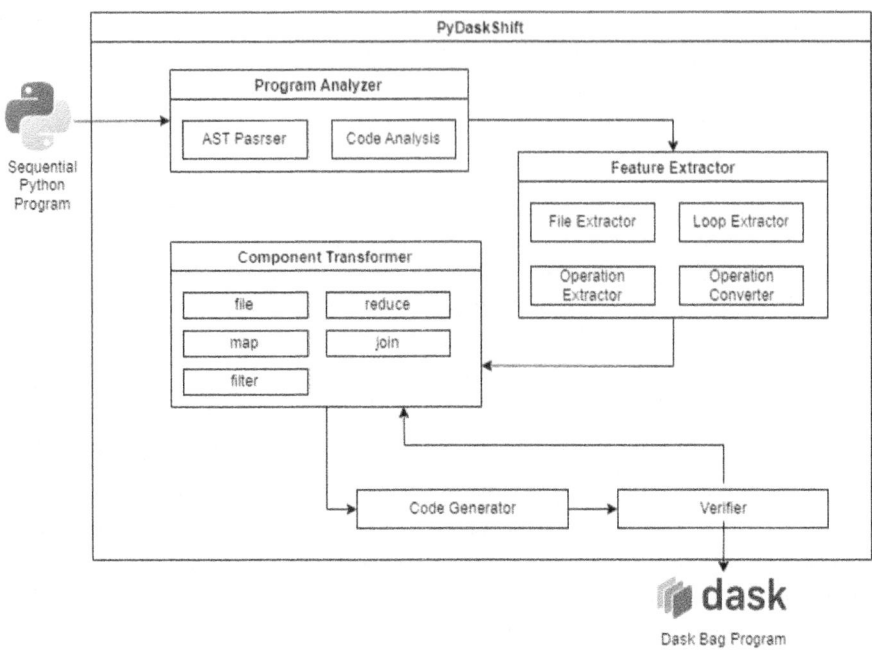

Fig. 1. The Software accepts Python programs as input. Then, it passes through different stages and produces the equivalent Dask Bag program.

```
1   import time
2   import csv
3   import sys
4
5   def minNum(numbers):
6       min1 = sys.maxsize
7       for n in numbers:
8           if n < min1:
9               min1 = n
10      return min1
11
12  def run_prog():
13
14      number = []
15      with open('minnumber.csv', newline='') as f:
16          reader = csv.reader(f)
17          for row in reader:
18              row = [ int(i) for i in row]
19              number += row
20      result = minNum(number)
21
22  if __name__ == '__main__':
23      run_prog()
```

Fig. 2. The program reads numbers from a file and stores them in a list. The *minNum* function iterates over the list, finds the minimum number, and returns it.

- File Transformer (Sect. 3.2)
- Filter Transformer (Sect. 3.3)
- Map Transformer (Sect. 3.4)
- Reduce Transformer (Sect. 3.5)

– Join Transformer (Sect. 3.6)
– UDF Transformer (Sect. 3.7)

Normally, PyDaskShift starts with the File Trasformer. The File Trans-
former transforms the file operations into Dask Bag operations from the meta-
information. Then, the Map Transformer starts its execution. However, if the
Nested Loop flag or the *UDF* flag is true, PyDaskShift begins with the Join
Transformer (Sect. 3.6) or the UDF Transformer (Sect. 3.7) respectively. The
Map Transformer transforms the extracted operations into Dask Bag map operations. The Code Generator then uses these map operations to generate the Dask
Bag program. If the generated program passes the user-provided test cases, the
code generation is complete.

However, the synthesis phases have inter-related functions. The Map or the
Reduce Transformer can employ the Filter Transformer to obtain all the filter
components of the current loop body. Similarly, the UDF Transformer can utilize
the Map Transformer or the Reduce Transformer to convert any operation into
a Dask Bag map or reduce operation.

3.2   File Transformer

The *FileInfo* class retains meta-information regarding file-read operations of the
Python program in Fig. 2 as follows:

– Start Line Number: *15*
– End Line Number: *19*
– File Name: *minnumber.csv*
– Target: *number*
– Data Type: *int*

PyDaskShift transforms the file-read operations by utilizing the *read_text* API
and splitting the values based on the comma delimiter (Fig. 3). It assumes the
file contains only numeric data and is in a *csv* format. The user provides the
blocksize parameter. This operation operates lazily, postponing the reading from
the file until the execution is explicitly initiated within the program.

In Fig. 3, Line 1 generates a Bag from the file. Dask will read this file in
parallel, generating a Dask Bag. The *blocksize* parameter determines the amount
of data read at once by Dask. Line 2 uses Dask's map function to transform each
element by applying a lambda function using the comma delimiter. The data type
int is employed to transform each Bag element, information that can be found
in the meta-information. Finally, Line 3 converts to a single Bag of integers.

```
1   number = daskbag.read_text('minnumber.csv',blocksize=blocksize).str.strip()
2   number = number.map(lambda x: [int(num) for num in x.split(',')])
3   number = number.flatten()
```

Fig. 3. Create Dask Bag from file with PyDaskShift execution

3.3   Filter Transformer

Before the execution of the Map Transformer, PyDaskShift checks the filter flag. If it is set, as is the case in our example by the *if* statement in Line 8, priority will be given to the Filter Transformer. The Filter Transformer examines the filter condition specified in the meta-information.

Subsequently, the Filter Transformer translates the filter condition by applying a lambda expression to the iteration variable n. In this case, the lambda expression *lambda n: n < min* fits the filter. This function is then enclosed within the Dask Bag API filter function as *filter(lambda n: n < min)*.

Presently, PyDaskShift only supports a single condition within a filter, but future plans call for the development of more complex statements. Any map or reduce operation contained within an *if* statement will be encompassed by the *filter* operation.

3.4   Map Transformer

The Map transformer transforms each loop operation into a Dask Bag map operation, generating a new Bag for each operation. In instances where multiple operations are present, each undergoes a similar conversion into a map operation. The *MapperList* stores the newly created Bags.

In our example scenario, the Map Transformer initiates its execution upon being invoked by the Filter Transformer. It proceeds by transforming each filter operation into map operations, generating a new Bag. Initially, the assign operation from the meta-information is encapsulated within a lambda function, nested inside the map operation information. Consequently, the generated map function appears as *map(lambda x: n)*. The map operation is then appended to the filter function from the Filter Transformer, completing the map transformation as *filter(lambda n: n < min).map(lambda x: n)*.

3.5   Reduce Transformer

If the map transformation verification fails, PyDaskShift moves on to the subsequent stage of synthesis: the Reduce Transformer. The Reduce function, distinct from a Bag, yields an aggregated value. The reduce transformer is similar to the map transformer, except the operation is enclosed within the syntax of the reduce function. Notably, Dask Bag allows for the use of the reduce function without requiring a map function.

In our example scenario, the outcome of the Map Transformer fails to satisfy the test cases. In particular, the map results would return one minimum value for each bag. Consequently, the Reduce Transformer commences the transformation into a reduce function. The lambda expression is nested within the *fold* operation and subsequently appended to the filter function generated by the Filter Transformer. The third row of Table 1 illustrates the resultant reduce operation converted by the Reduce Transformer. The reduce expression is further matched using pattern matching rules. Upon matching with the *min* function, the Reduce Transformer simplifies the reduce expression into a straightforward *min* function. PyDaskShift utilizes a pattern-matching rule to implement diverse reduction methods, as showcased in Table 1.

Table 1. PyDaskShift reduce conversion using pattern matching rule.

Transformed Operations	Dask Bag Operation
fold(lambda x,n:x+n)	sum()
fold(lambda x,n:x+1)	count()
filter(lambda n: n < min).fold(lambda n : n)	min()
filter(lambda n: n > max).fold(lambda n : n)	max()

However, a fundamental difference between map and reduce lies in the potential presence of a cumulative and associative value known as an accumulator. This accumulator's existence can be determined with the assistance of meta-information. If the target value matches either the left or right value, PyDaskShift identifies it as an accumulator. Consider the operation $sum = sum + n$, where n is an element of a list. In this case, the variable *sum* serves as the accumulator since the target and the left value are the same for binary operator.

- target: *sum*
- left: *sum*
- ops: *+*
- right: *n*

In this scenario, PyDaskShift transforms the function into a lambda function as follows: *lambda(sum, n : sum + n)*. This lambda function is then enclosed within the reduce syntax as *fold(lambda(sum, n : sum + n))*. *Fold* in Dask Bag is the built-in function for *reduce*.

3.6 Join Transformer

If the *Nested Loop* flag is true, the transformation starts with the Join Transformer. We assume that the nested loop comprises only two loops, with the operations situated within the inner loop. We also assume that the nested loops are join operations. Currently, PyDaskShift exclusively deals with inner-join and cross-product operations. If the inner loop contains an *append* operation, it is assumed that a possible join operation exists. Additionally, if there is an *if* statement, it is assumed that the join operation is an inner join.

In our example, there is no nested loop present. We use Fig. 4 to demonstrate a straightforward inner join and showcase the corresponding Dask Bag program produced by PyDaskShift. It specifically exhibits the key operations within the Dask Bag program.

The Join Transformer employs the Dask Bag *product* API to perform a cross join. To handle the absence of a built-in API for the inner join in Dask, PyDaskShift uses the filter operation to convert the inner join. When detecting the join, it manages the operations within the inner loop similarly to the earlier synthesis stages. Upon encountering a filter operation, the Filter Transformer modifies the filter condition. Subsequently, the operations within the *if* statement undergo transformation by the Map Transformer. In Fig. 4 (Line 5), the Join Transformer initially processes the join operation, followed by the transformation of filter operations by the Filter Transformer in Line 6. Finally, the Map Transformer transforms the map operations, which are then appended to the filter function.

```
for i in data_1:
        for j in data_2:
            if i[0] == j[0]:
                    data_3.append((i[0] , i[1], j[1]))


data_1_bag_result = data_1_bag.product(data_2_bag)
data_2_bag_result = data_1_bag_result.filter(lambda x: x[0][0] == x[1][0])\
                        .map(lambda x: (x[0][0], x[0][1] , x[1][1]))
```

Fig. 4. Python program for inner join and the corresponding Dask Bag program converted by PyDaskShift.

3.7   UDF Transformer

If the *UDF* flag is true, the UDF Transformer takes precedence over the Map Transformer. The UDF Transformer ensures parallelization of the function call inside the loop using the Dask Bag operation. Similar to the Filter Transformer, it utilizes the Map Transformer and, if the test case verification fails, employs the Reduce Transformer.

In our current example, there is no UDF call present. To illustrate the UDF Transformer's functionality, let us consider a different scenario depicted in Fig. 5. Here, the *udfexample* function involves a UDF call within the loop, and the UDF flag is set to true. The UDF Transformer identifies this information in the meta-information and uses the Map Transformer to modify the operation. If the verification fails, it proceeds to the next stage of gradual synthesis.

The map function encapsulates the UDF call, and the Map Transformer works in a similar way for the UDF call, wrapping the operation with the lambda expression *lambda x: square(x)* and nesting it within the map function as *map(lambda x: square(x))*.

```
def square(x):
    x = x*2
    return x

def udfexample(numbers):
    output = []
    for n in numbers:
        output.append(square(n))
    return output
```

Fig. 5. Python program example for UDF call.

If the converted operation passes the test cases, the conversion concludes. Otherwise, it continues through the other synthesis stages.

3.8 Code Generator

The Code Generator produces the final code as part of the conversion process. The Component Transformer provides the list of transformed operations based on the synthesis stage. First, the Code Generator add *import* statement required for the Dask Bag library. It also introduces two global variables: *workers* and *blocksize*. The workers variable is the number of Dask workers for execution [6]. The blocksize parameter in Dask controls the chunk size of data read from files. PyDaskShift accepts these two values as input, with the default value for *workers* set to 36 and the default value for *blocksize* established as 256 MB.

The replacement code is different based on the synthesis stage, For example, the replacement code could be from Table 1. Upon completion of replacing the stored list of map transformations within the MapperList, the Code Generator identifies the conclusion of the transformation process. Once all operations have been replaced, the Dask Execution activates by appending the *compute* function to the Dask Bag.

4 Results and Analysis

This chapter contains the findings from our performance analysis of both distributed and parallel Dask Bag programs in comparison to the initial sequential Python program for data processing. Our research aims to evaluate the Dask Bag-based approach's performance in terms of execution time while dealing with large datasets.

4.1 Experimental Setup

We evaluated PyDaskShift through a series of experiments, utilizing four AWS machines. These machines featured Ubuntu 22.04 OS, Intel(R) Xeon(R) CPU E5-2676 v3 @ 2.40 GHz, 1-Core processors, and 1GB of RAM each. For every experiment, we performed multiple runs and determined the average time (in seconds) for comparison. One machine executed the sequential program, while the distributed parallel Dask Bag program utilized all four machines. We categorized the sequential programs into specific groups.

– Aggregate Functions
– Join Operations
– K-Nearest Neighbors (KNN)
– Pi Calculation

In aggregate functions, we conducted computations encompassing metrics such as count, sum, minimum, maximum, and average. These metrics were applied to a given list of values, forming an integral part of our analysis. During our examination, we delved into inner join operation.

For our subsequent experimentation, we focused on two specific programs, namely the K-Nearest Neighbors (KNN) program and the pi calculation program.

4.2   Aggregate Functions

After running the programs with small datasets, we observed that, in such cases, the sequential Python programs outperform the parallel execution provided by Dask. Consequently, we scaled up the dataset size. Through multiple experiment iterations, we obtained average execution times, enabling a comprehensive comparison of performance. To ensure enhanced performance, we conducted our experiments using a file size of 1 GB.

In the context of computational tasks, the performance gain is a pivotal factor that signifies the efficiency enhancements introduced using alternative methods or tools. Table 2 functions as a graphical depiction of the enhanced performance, highlighting the benefits of utilizing PyDaskShift to transform the program into a Dask Bag program, as opposed to the traditional sequential approach.

Table 2. Speed up comparison of generated programs for the aggregated functions

Aggregate Functions	Sequential Program	Dask Bag Program	Speed Up
Sum	233	142	1.65x
Count	232	141	1.63x
Minimum	216	141	1.53x
Maximum	214	140	1.53x
Average	260	152	1.71x

Through careful observation and measurement of execution times, Table 2 highlights the reduction in processing time achieved by utilizing the Dask Bag program in contrast to the sequential program. This quantifiable difference underscores the benefits of leveraging Dask's parallel computing capabilities to optimize the execution of tasks, resulting in more efficient and expedient data processing workflows.

4.3   Join Operations

Join Operations are characterized by a nested loop structure. The program systematically traverses every element within the first list for each corresponding element within the second list. In an inner join, the program identifies instances where the first element of both lists matches, subsequently appending the corresponding data to the resulting set.

In our conducted experiments, we employed two files, each with a size of 1 KB. The execution time results for both sequential and Dask Bag implementations of join operations, along with their respective speedups, are presented in Table 3. This improvement is likely due to efficient implementation of the join operation within Dask.

Table 3. Speed up comparison of generated programs for the inner join and cross product

Join Operations	Sequential Program	Dask Bag Program	Speed Up
Inner join	6	4	1.5x

4.4   K-Nearest Neighbors (KNN)

K-Nearest Neighbors (KNN) is a machine learning algorithm for classification and regression tasks. It is a non-parametric instance-based learning method that makes predictions based on the similarity of input data points to the training data. In KNN, the K refers to the number of nearest neighbours considered for making a prediction.

The initial step of the program involves extracting data from a designated file. Subsequently, the program calculates the distance between each dataset entry and a specified point of interest. This distance computation serves to quantify the dissimilarity or similarity between the data points and the target point. The program sorts the dataset entries in ascending distance order. It returns the predetermined number of the dataset entries with the smallest distance from the specified point.

Parallelization of the computation of the distance function represents an optimization strategy that can significantly enhance the efficiency of the process [10]. Nevertheless, in our experimentation involving an 1 GB file and a value of k = 10, an intriguing outcome emerged (Table 4).

Table 4. Speed up comparison of the KNN and modified KNN program

KNN program	Sequential Program	Dask Bag Program	Speed Up
KNN	122	32	3.81x

```
import dask.bag as daskbag
from dask.distributed import Client
workers=36
blocksize = '256MB'

from math import sqrt
import csv

def euclidean_distance(row1, row2):
    distance = (row1[0] - row2[0]) ** 2 + (row1[1] - row2[1])**2
    return row1,sqrt(distance)

def get_neighbors(train, test_row):
    distances = list()
    with Client(n_workers=workers) as client:
        dist_bag_1 = train.map(euclidean_distance,test_row)
        dist = dist_bag_1.topk(10, key=lambda x: -x[1]).compute(num_workers=workers)
    return dist

def KNN(data, query, k):
    data_list = get_neighbors(data,query)
    return data_list

def run_prog():
    dataset = []
    dataset = daskbag.read_text('KNNdata.csv',blocksize=blocksize).str.strip()
    dataset = dataset.map(lambda x: [float(num) for num in x.split(',')])

    result = KNN(dataset,[3.2,1.2],10)

if __name__ == '__main__':
    run_prog()
```

Fig. 6. Generated Dask Bag program for the modified K-Nearest Neighbors (KNN)

The underlying assumption guiding this endeavour is that incorporating a sorting mechanism will enable the utilization of the *topk* method. This adaptation results in the structured form depicted in Fig. 6.

It is important to note that this strategic transformation holds the potential for broader applicability beyond the immediate context. The process employed here can be dissected and examined for its adaptability to various programs, serving as an avenue for potential future exploration and development. Implementing the Dask Bag program resulted in a 3.81-fold increase in execution speed..

4.5   Pi Calculation

Calculating the value of π through the Leibniz formula involves utilizing a mathematical series that gradually converges towards the value of π. The Leibniz formula, credited to the German mathematician Gottfried Wilhelm Leibniz, represents π as an alternating series:

$$\pi/4 = 1 - \frac{1}{3} + \frac{1}{5} - \frac{1}{7} + \frac{1}{9} - \ldots$$

In this expression, π is approximated by systematically adding and subtracting successive fractions. Including additional terms in the series improves the accuracy of this approximation. It is important to note that the Leibniz formula's convergence is relatively slow, necessitating many terms to achieve a high degree of precision.

PyDaskShift recognizes a potential avenue for transforming the loop segment contained within the *calculate_pi_leibniz* function. Our experimentation encompassed varying iterations to evaluate the performance of the approach comprehensively.

It is noteworthy that the precision of the computed pi value is contingent on the iteration count. Generally, an increased number of iterations results in a more accurate approximation of pi.

However, a distinctive observation emerged concerning the accuracy of the pi value in the Dask Bag program generated through our approach. Surprisingly, the accuracy does not necessarily augment with escalating iterations. A nuanced technical aspect underpins this phenomenon: within each worker, the computation entails the summation of values susceptible to floating-point precision errors. These errors can accumulate across the workers as they converge to produce the final result. As a result, the increased iteration count intensifies parallelism and speed. However, it does not proportionally enhance the precision of the pi value, due to the cumulative nature of these errors.

Table 5. Speed up comparison of the pi program

No. of iterations	Sequential Program	Dask Bag Program	Speed Up
1,000,000,000	182	157	1.16x

An intriguing observation during the conversion process pertains to the choice between using two different Bag constructors – *daskbag.from_sequence(range(iterations))* and *daskbag.range(iterations)* for parallelization. Initially, we encountered a situation where the former method resulted in longer execution times than the sequential program. However, upon adopting the latter approach, marked performance improvements were achieved for this program, as illustrated in Table 5. As a result, this new type of Bag creation was addded to the synthesis.

The rationale behind this distinction is rooted in the mechanics of these Dask Bag APIs. In the first method, *daskbag.from_sequence(range(iterations), npartitions = 6)*, the process involves first generating a range of integers and subsequently constructing a Bag from this range. This initial range generation step consumes a significant amount of time, impacting the overall efficiency of the program.

The pivotal change that yielded enhanced performance was using the Dask Bag API *daskbag.range (iterations, npartitions = 36)*. This method inherently generates the specified range of integers directly within the bag, bypassing the separate range creation step. Therefore, mitigating the overhead associated with creating the range leads to notable gains in execution speed.

Nevertheless, we can alleviate the inherent distinction between the two approaches using the Dask library, indicating a potential avenue for future enhancements. The observed internal disparity between the two methods underscores the underlying complexity of parallel computing and the intricacies of data handling and optimization. The Dask library, known for its capabilities in handling parallel and distributed computations, can potentially play a role in addressing this discrepancy.

4.6 Parallel Environment Experimentation

We also reproduced our experiments within a parallel environment on a single machine. We conducted our trials on a Linux system equipped with an Intel(R) Xeon(R) CPU E5-2695 v4 @ 2.10 GHz, featuring 36 cores, and 256 GB of RAM. However, for aggregated functions, we explored a file size of 9.3 GB, while for KNN, we utilized a file size of 1 GB. Additionally, for the inner join operation, we worked with a file size of 953 MB. The results of these experiments are summarized in Table 6.

Table 6. Speed up comparison of generated programs executed in the single machine

Aggregate Functions	Sequential Program	Dask Bag Program	Speed Up
Sum	1,820	267	6.8x
Count	1,932	278	6.94x
Minimum	1,810	268	6.75x
Maximum	1,802	265	6.8x
Average	1,821	277	6.57x
Inner join	252	53	4.75x
KNN	1,158	63	18x
Pi Caculation	285	67	4.25x

A comparative analysis of the speedup achieved in join operations and aggregate functions reveals that join operations do not exhibit the same level of rapid acceleration as aggregate functions. The data transfer overhead [4] arises from the necessity to share and exchange data between the cores to perform the join operation accurately. This communication between cores introduces additional computational costs and time, impacting the overall performance of the operation. In the case of KNN, there is a notable enhancement observed compared to the distributed environment. This improvement can be attributed to the utilization of 36-core machines, enabling the parallel execution of tasks across all cores simultaneously. As a result, the execution time is significantly reduced due to the efficient utilization of computational resources.

Due to the enhanced processing capabilities of the machine, our findings generally indicate significantly improved performance compared to a distributed environment. The robust computing power of the system allows for more efficient execution of tasks. Moreover, conducting experiments on a single machine minimizes data transfer across the machines, thereby leading to decreased execution times.

5   Future Work and Conclusion

PyDaskShift introduces a new tool that automatically transforms loop-based sequential Python programs into their corresponding Dask Bag API programs. This tool holds the distinction of being the first of its kind to execute such conversions, underscoring its pivotal role in advancing the automation of parallel programming in Python.

Employing a systematic approach, PyDaskShift utilizes compiler techniques and Gradual Synthesis to accomplish this conversion. It initially analyzes the structure, functions, and loops of the program using Python's AST. Subsequently, it extracts operations from the AST and converts them into the Dask Bag program with the assistance of gradual synthesis and operations. Following the completion of the code generation phase, PyDaskShift verifies the correctness of the translated program using the user-provided test cases. In the event of test case failure, the tool proceeds to the next stage of gradual synthesis, iteratively refining the program to achieve equivalence with the original sequential program.

PyDaskShift's capabilities offer numerous advantages for Python developers, significantly reducing the time and effort required to parallelize sequential Python programs. This makes it more accessible for developers to leverage the benefits of parallel computing. PyDaskShift conducted experiments with various programs, converting them into equivalent Dask Bag programs and achieving up to an 18x speed improvement compared to the original Python programs.

Currently PyDaskshift supports only two nested loops. Handling nested loops with more than two levels is a challenging task. To efficiently address this challenge, we identify the nested loop structure, break it into manageable components, and gradually parallelize one inner loop at a time, starting with the innermost loop. This gradual synthesis approach simplifies the analysis and transformation process while ensuring that each nested loop retains its original functionality. We apply pattern-matching rules and heuristics to identify parallelization opportunities, and we carefully manage data dependencies, especially in variables shared across nested levels, to avoid conflicts. The process involves iterative parallelization, comprehensive testing using various test cases, and performance evaluation on different datasets and hardware configurations. Handling diverse loop types and iterating to refine the parallelization are also essential considerations when dealing with nested loops with multiple levels.

In its current state, PyDaskShift does not optimize parameters for the generated Dask Bag program. We can enhance optimization by introducing an additional module dedicated to fine-tuning these parameters. This module should be capable of identifying the most efficient program in terms of execution time when the tool generates multiple programs. Moreover, it can suggest the ideal number of workers necessary for program execution and recommend the appropriate block size for file partitioning within the Dask Bag.

While PyDaskShift currently supports CSV files containing numbers, it can be expanded to handle various other file formats. This extension will require significant modifications to the File Transformer module. However, once a Dask Bag is created from the file, we anticipate only minor modifications to the rest of the synthesis process.

Furthermore, PyDaskShift currently focuses on converting programs to equivalent Dask Bag programs. However, we can generalize our Component Transformer module to other parallel programming framework and identify the most suitable program for specific problems. Later on, we can merge Tyro and PyDaskShift into a unified tool, allowing users to choose the output parallel programming framework. This extended functionality can serve as a recommendation system, helping users determine the best framework for parallelization in a given program.

The current state of PyDaskShift's reduce operation is confined to the Dask Bag *fold* API.However, there exist several other Dask Bag APIs, such as *foldby* and *groupby*, which can potentially be employed for reduction tasks. To implement this broader functionality, we must adjust the Component Transformer module. In doing so, we also need to expand our search space to accommodate these additional APIs effectively. This expansion will allow PyDaskShift to leverage a broader range of reduction techniques and improve its capacity to handle diverse reduction operations in a more versatile manner.

At present, our evaluation of PyDaskShift involved scrutinizing its performance across 9 distinct programs (Sect. 4.1). To broaden our assessment, we wanto to increase the number of programs under scrutiny. Specifically, we plan to introduce additional programs such as conditional sum, conditional count, and matrix multiplication into our evaluation criteria. By including these tasks, we aim to explore PyDaskShift's efficiency and effectiveness in handling a wider range of operations. This expansion in evaluation criteria will provide a more comprehensive understanding of PyDaskShift's capabilities and performance across various types of computational workloads.

References

1. Islam, A.: Pydaskshift: converts sequential python program to distributed parallel program. https://github.com/agmislam1/parallelpy/tree/main/
2. Ahmad, M.B.S., Cheung, A.: Automatically leveraging MapReduce frameworks for data-intensive applications. In: Proceedings of the 2018 International Conference on Management of Data, SIGMOD 2018, pp. 1205–1220. Association for Computing Machinery, New York (2018). https://doi.org/10.1145/3183713.3196891
3. Boyer, R.S., Moore, J.S.: A mechanical proof of the unsolvability of the halting problem. J. ACM **31**(3), 441–458 (1984). https://doi.org/10.1145/828.1882

4. Böhm, S., Beránek, J.: Runtime vs scheduler: analyzing Dask's overheads. In: 2020 IEEE/ACM Workflows in Support of Large-Scale Science (WORKS), pp. 1–8 (2020). https://doi.org/10.1109/WORKS51914.2020.00006
5. Ciccozzi, F., Addazi, L., Asadollah, S.A., Lisper, B., Masud, A.N., Mubeen, S.: A comprehensive exploration of languages for parallel computing. ACM Comput. Surv. **55**(2) (2022). https://doi.org/10.1145/3485008
6. Álvarez Cid-Fuentes, J., Álvarez, P., Solà, S., Ishii, K., Morizawa, R.K., Badia, R.M.: ds-array: a distributed data structure for large scale machine learning (2021)
7. Culler, D., et al.: LogP: towards a realistic model of parallel computation. SIGPLAN Not. **28**(7), 1–12 (1993). https://doi.org/10.1145/173284.155333
8. Dask Development Team: Dask documentation (2023). https://docs.dask.org/en/latest/. Accessed 28 Aug 2023
9. Dean, J., Ghemawat, S.: MapReduce: simplified data processing on large clusters. Commun. ACM **51**(1), 107–113 (2008). https://doi.org/10.1145/1327452.1327492
10. Du, S., Li, J.: Parallel processing of improved KNN text classification algorithm based on Hadoop. In: 2019 7th International Conference on Information, Communication and Networks (ICICN), pp. 167–170 (2019). https://doi.org/10.1109/ICICN.2019.8834973
11. Fedyukovich, G., Ahmad, M.B.S., Bodik, R.: Gradual synthesis for static parallelization of single-pass array-processing programs. SIGPLAN Not. **52**(6), 572–585 (2017). https://doi.org/10.1145/3140587.3062382
12. Fegaras, L., Noor, H.: Translation of array-based loops to distributed data-parallel programs. Proc. VLDB Endow. **13**(8), 1248-1260 (2020). https://doi.org/10.14778/3389133.3389141
13. Frachtenberg, E., Schwiegelshohn, U.: New challenges of parallel job scheduling. In: Frachtenberg, E., Schwiegelshohn, U. (eds.) Job Scheduling Strategies for Parallel Processing, pp. 1–23. Springer, Heidelberg (2008). https://doi.org/10.1007/978-3-540-78699-3_1
14. Hoare, C.A.R.: An axiomatic basis for computer programming. Commun. ACM **12**(10), 576–580 (1969). https://doi.org/10.1145/363235.363259
15. Jiang, L., Su, Z.: Automatic mining of functionally equivalent code fragments via random testing. In: Proceedings of the Eighteenth International Symposium on Software Testing and Analysis, ISSTA 2009, pp. 81–92. Association for Computing Machinery, New York (2009). https://doi.org/10.1145/1572272.1572283
16. Katsifodimos, A., Schelter, S.: Apache Flink: stream analytics at scale. In: 2016 IEEE International Conference on Cloud Engineering Workshop (IC2EW), p. 193 (2016). https://doi.org/10.1109/IC2EW.2016.56
17. Manna, Z., Waldinger, R.: A deductive approach to program synthesis. ACM Trans. Program. Lang. Syst. **2**(1), 90–121 (1980). https://doi.org/10.1145/357084.357090
18. Meade, A., Buckley, J., Collins, J.J.: Challenges of evolving sequential to parallel code: an exploratory review. In: Proceedings of the 12th International Workshop on Principles of Software Evolution and the 7th Annual ERCIM Workshop on Software Evolution, IWPSE-EVOL 2011, pp. 1–5. Association for Computing Machinery, New York (2011).https://doi.org/10.1145/2024445.2024447
19. Pierre-Etienne, M., Ringeissen, C., Vittek, M.: A pattern matching compiler for multiple target languages. In: Hedin, G. (ed.) CC 2003. LNCS, pp. 61–76. Springer, Heidelberg (2003). https://doi.org/10.1007/3-540-36579-6_5
20. Sanjel, A.: Tyro: a first step towards automatically generating parallel programs from sequential programs. Ph.D. thesis, Baylor University (2020)
21. Sarkar, T.: Parallelized Data Science, pp. 257–298. Apress, Berkeley (2022). https://doi.org/10.1007/978-1-4842-8121-5_10

22. Shvachko, K., Kuang, H., Radia, S., Chansler, R.: The Hadoop distributed file system. In: 2010 IEEE 26th Symposium on Mass Storage Systems and Technologies (MSST), pp. 1–10 (2010). https://doi.org/10.1109/MSST.2010.5496972

23. Zaharia, M., et al.: Resilient distributed datasets: a fault-tolerant abstraction for in-memory cluster computing. In: 9th USENIX Symposium on Networked Systems Design and Implementation (NSDI 12), pp. 15–28. USENIX Association (2012). https://www.usenix.org/conference/nsdi12/technical-sessions/presentation/zaharia

Author Index

© The Editor(s) (if applicable) and The Author(s), under exclusive license
to Springer Nature Switzerland AG 2024
H. Han (Ed.): SDSC 2024, CCIS 2158, pp. 235–236, 2024.
https://doi.org/10.1007/978-3-031-67871-4

GPSR Compliance

The European Union's (EU) General Product Safety Regulation (GPSR) is a set of rules that requires consumer products to be safe and our obligations to ensure this.

If you have any concerns about our products, you can contact us on ProductSafety@springernature.com

In case Publisher is established outside the EU, the EU authorized representative is:

Springer Nature Customer Service Center GmbH
Europaplatz 3
69115 Heidelberg, Germany

The manufacturer's authorised representative in the EU is Springer
Nature Customer Service Centre GmbH, Europaplatz 3, 69115 Heidelberg,
Germany. If you have any concerns regarding our products, please
contact ProductSafety@springernature.com

Printed and bound by CPI Group (UK) Ltd, Croydon, CR0 4YY
05/05/2026
02103581-0001